AF828719

Planning Middle Eastern Cities

An urban kaleidoscope in a globalizing world

Planning, History and Environment Series

Editor:
Professor Dennis Hardy, Middlesex University, UK

Editorial Board:
Professor Arturo Almandoz, Universidad Simón Bolivar, Caracas, Venezuela
Professor Nezar AlSayyad, University of California, Berkeley, USA
Professor Eugenie L. Birch, University of Pennsylvania, Philadelphia, USA
Professor Robert Bruegmann, University of Illinois at Chicago, USA
Professor Jeffrey W. Cody, Chinese University of Hong Kong, Hong Kong
Professor Robert Freestone, University of New South Wales, Sydney, Australia
Professor David Gordon, Queen's University, Kingston, Ontario, Canada
Professor Sir Peter Hall, University College London, UK
Professor Peter Larkham, University of Central England, Birmingham, UK
Professor Anthony Sutcliffe, Nottingham University, UK

Technical Editor
Ann Rudkin, Alexandrine Press, Oxford, UK

Published titles

The Rise of Modern Urban Planning, 1800–1914 edited by Anthony Sutcliffe

Shaping an Urban World: Planning in the twentieth century edited by Gordon E. Cherry

Planning for Conservation: An international perspective edited by Roger Kain

Metropolis 1980–1940 edited by Anthony Sutcliffe

Arcadia for All: The legacy of a makeshift landscape by Dennis Hardy and Colin Ward

Planning and Urban Growth in Southern Europe edited by Martin Ward

Thomas Adams and the Modern Planning Movement: Britain, Canada and the United States by Michael Simpson

Holford: A study in architecture, planning and civic design by Gordon E. Cherry and Leith Penny

Goodnight Campers! The history of the British holiday camp by Colin Ward and Dennis Hardy

Model Housing: From the Great Exhibition to the Festival of Britain by S. Martin Gaskell

Two Centuries of American Planning edited by Daniel Schaffer

Planning and Urban Growth in the Nordic Countries edited by Thomas Hall

From Garden Cities to New Towns: Campaigning for town and country planning, 1899–1946 by Dennis Hardy

From New Towns to Green Politics: Campaigning for town and country planning 1946–1990 by Dennis Hardy

The Garden City: Past, present and future edited by Stephen V. Ward

The Place of Home: English domestic environments by Alison Ravetz with Richard Turkington

Prefabs: A history of the UK temporary housing programme by Brenda Vale

Planning the Great Metropolis: The 1929 Regional Plan of New York and Its Environs by David A. Johnson

Rural Change and Planning: England and Wales in the twentieth century by Gordon E. Cherry and Alan Rogers

Of Planting and Planning: The making of British colonial cities by Robert Home

Planning Europe's Capital Cities: Aspects of nineteenth-century urban development by Thomas Hall

Politics and Preservation: A policy history of the built heritage, 1882–1996 by John Delafons

Selling Places: The marketing and promotion of towns and cities, 1850–2000 by Stephen V. Ward

Changing Suburbs: Foundation, form and function edited by Richard Harris and Peter Larkham

The Australian Metropolis: A planning history edited by Stephen Hamnett and Robert Freestone

Utopian England: Community experiments 1900–1945 by Dennis Hardy

Urban Planning in a Changing World: The twentieth experience edited by Robert Freestone

Twentieth-Century Suburbs: A morphological approach by J.W.R. Whitehand and C.M.H. Carr

Council Housing and Culture: The history of a social experiment by Alison Ravetz

Planning Latin America's Capital Cities, 1850–1950 edited by Arturo Almandoz

Titles published in 2003

Exporting American Architecture, 1870–2000 by Jeffrey W. Cody

Planning by Consent: the Origins and Nature of British Development Control by Philip Booth

Titles published 2004

The Making and Selling of Post-Mao Beijing by Anne-Marie Broudehoux

Planning Middle Eastern Cities: An urban kaleidoscope in a globalizing world edited by Yasser Elsheshtawy

Planning Middle Eastern Cities

An urban kaleidoscope in a globalizing world

edited by

Yasser Elsheshtawy

Routledge
Taylor & Francis Group
LONDON AND NEW YORK

First published 2004 by Routledge,
11 New Fetter Lane,
London EC4P 4EE

Simultaneously published in the USA and Canada
by Routledge
29 West 35th Street, New York, NY 10001

Routledge is an imprint of the Taylor & Francis Group

© 2004 Selection and editorial material Yasser Elsheshtawy; individual chapters: the contributors

Typeset in Palatino and Humanist by PNR Design, Didcot, Oxfordshire
Printed and bound in Great Britain by St Edmundsbury Press, Bury St Edmunds, Suffolk

This book was commissioned and edited by Alexandrine Press, Marcham, Oxfordshire

All rights reserved. No part of this book may be reprinted or reproduced or utilised in any form or by any electronic, mechanical or other means, now known or hereafter invented, including photocopying and recording, or in any information storage or retrieval system, without permission in writing from the publishers.

The publisher makes no representation, express or implied, with regard to the accuracy of the information contained in this book and cannot accept any legal responsibility or liability for any errors or omissions that may be made.

British Library Cataloguing in Publication Data

A catalogue record of this book is available from the British Library

Library of Congress Cataloging in Publication Data

Planning Middle Eastern cities : an urban kaleidoscope in a globalizing world / edited by Yasser Elsheshtawy.
p. cm. — (Planning, history, and the environment series)
Includes bibliographical references and index.
ISBN 0-415-30400-8 (hb)
1. City planning—Middle East. I. Elsheshtawy, Yasser. II. Series

HT169.M628P58 2004
307.1′216′0956—dc22 2004005323

Contents

Preface

This book originated from a desire to articulate a local voice. Many writings on the Middle East are from external observers. While such efforts are commendable they may miss some of the nuances of our region which can only be detected by those who are – on a daily basis – dealing with the problems of the 'Arab' street. Whether it is in encounters with citizens and authorities or dealing with various public and private organizations, all play a vital role in constructing an understanding, a referential framework, which may be absent for those observing our cities from afar. While such closeness may lead to a subjective viewpoint – as opposed to the 'objective', 'neutral' external observer – it nevertheless contributes to an empathy and an understanding of the 'deep structure' of Middle East cities. It is of interest to note however that all writers – except one – have left their 'home' cities and have settled elsewhere in the Middle East after a period of studying in the West. They exemplify the condition of the 'modern nomad' moving from one place to another, not belonging to a fixed country. At the same time living in, and being exposed to, more than one Middle East city has led to a better comprehension of our urban environment, particularly in appreciating its heterogeneous nature. Furthermore, being exposed to a variety of cultures, religions, and languages has led to an openness, and an understanding of the 'other'. Rather than stereotyping cultures which are different from our own we are trying to utilize the methods and tools of the West to further our understanding of our own societies, and in many ways to engage in a period of self-reflection and criticism rather than denial.

Another motivation for this book is the desire to dismantle stereotypes: whether it is in how the Middle East city is conceptualized, as I argue in the introduction, or how scholars employed in Arab universities are sometimes viewed. All but two of the contributors are working in Arab scientific institutions. It is rare that a chance is offered for people in our region to write about issues of vital interest to us and directed at a Western audience. The widely cited Arab Human Development Report, commissioned by the United Nations, has acquired its relevance and significance from the fact that it was written by Arabs thus conferring legitimacy. We hope that such efforts would be replicated, and this book should be viewed in that context. Scholars and academics in the Arab world have a lot to offer – many have elected to 'return' to improve the condition of their region – recognizing that the only way out of our current state of underdevelopment is to learn from the West and to assess critically our own values and beliefs – a fact made more urgent in the post 9/11 era. These tragic events have projected on our culture a perception of intolerance of others, yet they have also shown that we need to re-examine our own societies and the extent to which we are still struggling with 'modernity'.

I would like to acknowledge the effort of the contributors in this book. Their dedication and enthusiasm has been admirable, especially in following a tight schedule and in responding to some of my comments. The difficulty in balancing their various obligations while working on their respective chapters is appreciated. Ann Rudkin, deserves a special recognition for her vital role in bringing this book about, and her diligent effort in editing and correcting the chapters. Many of the ideas present in this book were initiated, discussed and emerged in a series of encounters I had with my peers in various organizations. There are simply too many to mention them all by name. However I would like specifically to note my teacher Amos Rapoport whose dedication and professionalism were an inspiration to me and who helped me to grow intellectually. Among the various organizations I would like to mention is IASTE (International Association for the Study of Traditional Environments) whose conferences helped me in formulating some of my main themes and arguments. While working on this book I was invited by Harvard Design School to talk about Dubai in a conference devoted to this city. The intellectual discourse of this event helped me in crystallizing many constructs and draw conclusions pertaining to Middle East cities, thus I am deeply indebted to them – specifically Professor Hashim Sarkis and the enthusiastic group of students who organized this event (MEDINA – Middle East Design Initiative North America). Thanks are also due to Dubai Municipality and Dubai Executive Council for sponsoring this event. Finally, I would like to acknowledge the support of UAE University for creating an environment that encourages scientific research. In particular, the Scientific Research Council as well as my colleagues at the Department of Architecture deserve special mention in this regard. While we live in a remote corner of the world, which in many ways is conducive to reflection, globalization and its 'contents' enabled us to share and communicate ideas easily. Living at the edge, surrounded by an infinite desertscape, has its merits after all!

Yasser Elsheshtawy
Al-Ain
May 2004

Illustration Credits and Sources

The editor, contributors and publisher would like to thank all those who have granted permission to reproduce illustrations. We have made every effort to contact and acknowledge copyright holders, but if any errors have been made we would be happy to correct them at a later printing.

Chapter 1

1.1.–1.5. © Yasser Elsheshtawy

Chapter 3

3.1. *Source*: Osmani, M. (1988) Physical Form and Spatial Structure of Colonial Cities in Algeria. Unpublished MPhil Thesis, Joint Centre for Urban Design, Oxford Brookes University, Oxford.
3.2. © Mohamed Osmani
3.3. *Source*: Osmani, M. (1988) Physical Form and Spatial Structure of Colonial Cities in Algeria. Unpublished MPhil Thesis, Joint Centre for Urban Design, Oxford Brookes University, Oxford
3.4. © Mohamed Osmani
3.5. *Source*: Osmani, M. (1988) Physical Form and Spatial Structure of Colonial Cities in Algeria. Unpublished MPhil Thesis, Joint Centre for Urban Design, Oxford Brookes University, Oxford
3.6. © Mohamed Osmani
3.7. © DigitalGlobe, www.digitalglobe.com
3.8. © Mohamed Osmani
3.9.–3.14. © Karim Hadjri
3.15. *After*: Institut National de Cartographie et de Télédétection (1999) *Alger: plan des rues*. Algiers: INCT and EPAU & SIAAL (2000) *Alger Métropole: Région – Ville – Quartier*. Ecole Polytechnique d'Architecture et d'Urbanisme (EPAU), Alger in cooperation with Städtebau-Institut, Stuttgart University (SIAAL), Germany, and Deutsche Gesellschaft für Technische Zusammenarbeit GmbH (GTZ), Eschborn, Germany

Chapter 4

4.1. *After*: Lassner, Jacom (1970) *The Topology of Baghdad in Early Middle Ages*. Detroit: Wayne State University Press, p. 27.
4.2.–4.3. © Ministry of Housing, Baghdad
4.4.–4.5. © Rifat Chadiji
4.6. *Source*: Chadirji, Rifat (1991) *Al-Ukhaider and the Crystal Palace*. London: Ryad El-Rayyes Books Ltd, p. 405
4.7. *Source*: Makiya, Kanan (1990) *Post Islamic Classicism*. London: Saqi Books. p. 48
4.8.–4.9. © Hoshiar Nooraddin

Chapter 5

5.1. *Photo*: Daniel Pini, by courtesy of the General Organization for the Preservation of Historic Cities of Yemen
5.2. *After*: Manzoni, R. (1884) *El Yemen*. Rome.
5.3. *After:* Lane, M. B. (1988) *Sana'a: Pilot Restoration Projects*. UNDP/UNESCO YEM/88/006
5.4. © Khaled Al-Sallal
5.5. *Photo*: Daniel Pini, by courtesy of the General Organization for the Preservation of Historic Cities of Yemen
5.6. *After*: Matthews, D. (1979) Building in the Yemen. Report in the Proceedings of a Seminar Held at the Museum of Ethnography, Rotterdam, by courtesy of the British-Yemeni Society
5.7. © Khaled Al-Sallal
5.8. *After*: HABITAT (2001) *Cities in A Globalizing World, Global Report on Human Settlement*. Nairobi: United Nations Centre for Human Settlements
5.9. *After*: Dresch, P. (2000) *A History of Modern Yemen*. Cambridge: Cambridge University Press
5.10. *Source*: General Organization for the Preservation of Historic Cities of Yemen

5.11. © Khaled Al-Sallal
5.12. *Source*: Al-Sallal, K.A. (1996) Solar access/shading and building form: geometrical study of the traditional housing cluster in Sana'a, in Sayigh A. (ed.) *Proceedings of the World Renewable Energy Congress*, Vol. 3. Oxford: Pergamon-Elsevier.
5.13.–5.14. © Khaled Al-Sallal
5.15. *Photo*: Daniel Pini, by courtesy of the General Organization for the Preservation of Historic Cities of Yemen
5.16. © Khaled Al-Sallal

Chapter 6
6.1. *After*: Brown, Carl, L. (ed.) (1973) *From Madina to Metropolis: Heritage change in the Near Eastern City*. Princeton, NJ: The Darwin Press
6.2. *After*: Abdelkafi, J. (1989) *La Médina de Tunis*. Paris: CNRS; Tunis: Alif
6.3. Master Plan of North Lake of Tunis (after the SPLT plan).
6.4.–6.5. © Bechir Kenzari
6.6.–6.7. *After*: Société de Promotion du Lac de Tunis (SPLT) (1998) *Plan d'Aménagement Général*. Tunis: SPLT.
6.8. *After*: Van Berk A. H. and Oostinga H. (1992) North Lake of Tunis and its shores: restoration and development. *Terra et Aqua, International Journal on Public Works, Ports and Warer ways Developments*, No. 49.

Chapter 7
7.1.–7.6. © Elke Vogel-Adham

Chapter 8
8.1.–8.8 © Yasser Elsheshtawy
8.9. © www.digitalglobe.com
8.10. © DigitalGlobe, www.digitalglobe.com
8.11.–8.15. © Yasser Elsheshtawy

Contributors

Khaled Adham is Assistant Professor of Architecture at the Suez Canal University, Egypt. He was born in Cairo in 1964. After receiving a bachelor degree from Cairo University in 1986, he worked for architects Rami El-Dahan (a student of Hassan Fathy) and Abdel Halim Ibrahim (CDC). In 1992 he received a Masters degree in Architecture from Kent State University and in 1997 a PhD in architecture from Texas A&M University. Since 1998, he has been teaching architectural theory and design studios at the Suez Canal University as well as practising architecture and urban planning in his private studio. His current research focuses on the late twentieth century architectural and urban transformations of Cairo.

Jamel Akbar is a Professor in the School of Architecture and Planning, King Faisal University, Dammam, Saudi Arabia. He holds a BArch from King Saud University, Riyadh and MArchAS and PhD from the MIT, USA. He has several publications in Arabic and English, among which are two books: *Crisis in the Built Environment: the Case of the Muslim City* (1988) and *'Imarat al-'ard fi al-'islam* (*The Architecture of the Earth in Islam*) (1992). In 1986 he was awarded the King Fahd Award for Architectural Research in the Muslim World.

Khaled A. Al-Sallal was born in Sana'a in 1960, and raised in Yemen as well as Egypt. After obtaining a Bachelor of Architecture degree from Ain-Shams University, Cairo in 1982, he went back to Yemen where he practiced architecture for almost a decade. He received a Master of Environmental Planning from Arizona State University in 1988 and a PhD from Texas A&M University in 1995. He has been teaching architecture at the UAE University in Al-Ain since 1996. His research has focused on vernacular and climatic responsive architecture in Yemen and the Arabian Gulf as well as energy and daylighting design in buildings.

Yasser Elsheshtawy was born in Cairo in 1963, raised in Switzerland and Germany and eventually graduated from the German Evangelical High School in Cairo. After he obtained a Bachelor of Architecture degree from Cairo University in 1986, he received a Masters Degree in Architecture from Pennsylvania State University in 1991 and a PhD from the University of Wisconsin-Milwaukee in 1996. He has been teaching architecture at the UAE University in Al-Ain since 1997. His research has focused on environment-behaviour studies, architectural theory as well as changing urban/architectural patterns of Middle Eastern cities.

Karim Hadjri was born in Algeria in 1961. He obtained a Diploma in Architecture from University of Sciences and Technology of Oran, Algeria in 1985. He was awarded a MPhil in 1989 and a PhD in 1992 from the Joint Centre for Urban Design at Oxford Brookes University in Oxford. He has worked as a scholar in the United Kingdom and Saudi Arabia, and managed academic units and research centres in both Cyprus and Colombia. Currently he teaches architecture at UAE University. His research interests are concerned with the design process, the use of Computer Aided Design, and applications of Geographic Information Systems in architecture, urban design and heritage management.

Bechir Kenzari was born in Tunisia in 1957 where he received his Masters Degree in Architecture in 1984 from the ITAAUT. He obtained his PhD from Georgia Tech in 1991, then taught for seven years at the University in Auckland, New Zealand. He is currently teaching architecture at the Department of Architectural Engineering in Al-Ain, UAE. He has an interest in linguistics, philosophy, fine arts and fiction. He published several short stories in the Auckland-based magazine *The Pander*, and has produced a good collection of oil paintings. His hobbies include deep-sea fishing.

Hoshiar Nooraddin was born in 1961 in the Iraqi Kurdish town Erbil. He studied architecture at the University of Technology in Baghdad, and obtained a PhD in Architecture from the Norwegian University of Science and Technology (NTNU) in Trondheim in 1996. He has professional experience in several countries and won a number of awards among them a first competition prize for Salahaddin University Headquarters (UNESCO-Erbil Iraq) in 2002. He is involved in activities related to environment, culture, democracy and human rights and currently works as a Senior Architect at the Urban Design Department, Norwegian Public Roads Administration in Oslo. He is also involved at the present with a reconstruction project in Iraq.

Mohamed Osmani graduated in Architecture from University of Sciences and Technology of Oran, Algeria in 1985. On completion of his MPhil at the Joint Centre for Urban Design at Oxford Brookes University in Oxford in 1988, he joined the Hubbard Ford Partnership in Bristol, UK, as an architect and was responsible for the planning, design and administration of a wide range of architectural and urban design projects. This was followed by a period as a Lecturer in Architecture and Computer Aided Architectural Design in Saudi Arabia. He currently teaches Architectural Engineering and Design Management at Loughborough University. His research focuses on sustainable design and construction, IT applications to building design and construction, Computer Aided Learning (CAL) and the built environment in developing countries.

Over here is like there, neither better nor worse. But I am from here, just as the date palm standing in the courtyard of our house has grown in *our* house and not in anyone else's. The fact that they came to our land, I know not why, does that mean that we should poison our present and our future? Sooner or later they will leave our country, just as many people throughout history left many countries. The railways, ships, hospitals, factories and schools will be ours and we'll speak their language without either a sense of guilt or a sense of gratitude. Once again we shall be as we were – ordinary people – and if we are lies we shall be lies of our own making.

Tayeb Saleh, *Season of Migration to the North*
(translated by Denys Johnson-Davies and published in 1970 by Heinemann)

Chapter 1

The Middle East City: Moving beyond the Narrative of Loss

Yasser Elsheshtawy

Was sind schon Städte, gebaut, ohne die Weisheit des Volkes.
(What are cities, built, without the wisdom of its citizens)

Bertolt Brecht, 1953

This book is a collection of chapters on what could be loosely referred to as 'Middle Eastern' or 'Arab/Moslem' cities. Both of these designations are placed in parentheses to demonstrate the rather vague nature of the terms and also their implications. Simply designating a city such as Cairo, for example, as 'Arab' or 'Moslem', associates a set of presumptions and preconceptions which may not be an accurate reflection of a city coping with nearly 20 million inhabitants, negotiating agreements with world trade organizations, and handling advances in information technology – to name just a few of the issues that may be present in any city throughout the world (figure 1.1). Such a perception may lead to forms of action which could be counterproductive, thus potentially contributing to the current state of underdevelopment present in many cities in the region. In using a geographical designation, which subsumes ethnic, religious and social groupings, a more accurate – and non-conventional – reading could be attained. The book thus differs from what has been, for the most part, a rather traditional reflection on cities inhabited by Arabs in the Middle Eastern region. With that in mind I will be using the terms 'Arab' and 'Middle Eastern' interchangeably.[1]

The six cities included here are not meant to be representative of all cities in the region but they are diverse enough to support the contention that there is no single monolithic entity termed 'Moslem/Arab'. They range from traditional centres (Cairo, Tunis, and Baghdad), 'fringe' cities (Sana'a and Algiers) to oil-rich Gulf entities (Dubai). And while this may be viewed as a general classification the distinction between cities in each category is quite broad thus even questioning the terms 'traditional', 'fringe' and 'oil-rich'. Cities within the Arab world are simply too diverse and hybrid to be lumped together under one 'arbitrary' category – hence the metaphor of the *kaleidoscope* used in the title. It suggests that there are multiple ways of looking at cities in our

Figure 1.1. Old Cairo: the traditional image of an 'Arab/Islamic' city.

region – similar to any other city in the world. Thus, like a kaleidoscope where with a quick turn of the hand patterns are re-arranged to form a new whole, these chapters represent a particular way of seeing, understanding, and representing the Middle Eastern city. In many ways such a characterization echoes James Donald's analogy of the city as a 'text' – as such meanings and symbols are ascribed to the physical environment, and it becomes a 'representation', an 'imagined environment' construed from the particular vantage point of each observer (Donald, 1992). Anthony King's edited collection on the twenty-first century metropolis is one example which applies such a framework (King, 1996). The Middle East city is curiously lagging behind. This book aims to correct this shortcoming.

To illustrate this point further I would like to discuss briefly a set of issues which will aid in understanding the structure of the book and the extent to which it enhances, and adds to what has been referred to as 'Middle Eastern Urban Studies'. First, I will review the existing literature on this subject which in my view revolves around, and is dominated by, a singular and short-sighted theme – namely a socio/religious reading of urban spatial patterns. This is particularly the case for Arab/Moslem writers. Second, I will explore the relationship of Arab cities to the current globalization discourse, to highlight both commonalities and differences with developments taking place at a global level. This discussion should, hopefully, provide a framework through which the chapters may be understood.

Finally, I will discuss the book's methodology in terms of initial questions asked of the contributors, emerging themes, and the structure of the book.

The Narrative of Loss

The typical architectural/urban narrative of the Middle Eastern city is a narrative of loss. A great, once flourishing civilization has through colonization been subjected to plundering and exploitation of resources – and thus cities are in a perpetual state of underdevelopment. Furthermore, the argument goes, this trend is exacerbated and amplified through the current universalization (or globalization) discourse. Traditional readings of the Middle Eastern city thus dominate, showing it as an isolated entity somehow disconnected from developments occurring elsewhere in the 'civilized' world. The city is always examined in relation to, and in association with, 'heritage', 'tradition', 'culture', as if it were divorced from the surrounding reality. Furthermore, the 'Islamic' perspective becomes a framework through which every single decision is evaluated, judged, and criticized. All developments taking place are framed within such an outlook, even in their interaction with 'modern' counterparts. Examples are numerous but perhaps the work of Al-Hathloul (1996) and Hakim (1986) could be considered representative in this regard, by attempting to establish a legal framework through which the Moslem city emerged and developed. Various decisions pertaining to the built environment are thus always referred back to religious texts. Conferences in the 1980s and 1990s further legitimized such approaches, focusing on such constructs as the 'Arab' city and the extent to which its 'glory' could be revived by tying it to its cultural/religious roots (for example, Serageldin and El-Sadek, 1982). Attempts at modernization and development are constructed *vis à vis* such a framework. It is as if one were to examine the city of London from a Western-Christian city model perspective – an absurd proposition yet somehow making sense within an 'Arab-Moslem' context. It is worth noting here that (European) colonialism did not just occur in the Middle East or other 'obvious' places but many other cities such as those in North America (New York and Washington for example). As Anthony King noted these cities have been insufficiently examined. Studying the Middle East city in relation to colonialism should thus be seen in the context of a 'larger colonial narrative' (King, 1996, p. 10). Such a conceptualization will inevitably lead to a realization that colonization may not be the sole factor responsible for the underdevelopment of the Middle East city.

This leads to a larger point: it is assumed that the 'Moslem' (Arab) has been unable to develop, grow, and in turn modernize – in short go beyond the twelfth century, the pinnacle of Moslem civilization. S/he is thus condemned to remain within this historic perspective. This, of course, has been the Orientalist reading of the Moslem/Arab mind initially exposed by Edward Said from a literary perspective (Said,1979) but examined through writers such as Zeynep Çelik dismantling the colonial French discourse in Algiers (Çelik, 1997), and Timothy Mitchell exploring colonial policies in Egypt, to cite a few (Mitchell, 1988). Shirine Hammadeh puts forward an interesting argument pertaining to the origins of the 'traditional' city concept which in her view comes from French colonial discourse as a way to dominate an unruly and exotic populace. The dual city construct – having both a 'traditional' settlement and a European one – is construed as a 'freezing of the image of a society in time and space' thus maintaining a physical differentiation between the colonizers and the colonized (Hamadeh, 1992). Continuing this trend Nelida Fuccaro in examining urban studies in the Gulf region calls for a comparative perspective both grounding urban settlements within their regional contexts and studying the 'specificity' of each – thus moving away from the 'static ideal type' of the 'Islamic city'. Her review of the literature shows that there is an emerging body of research which examines Gulf cities from a socio-political perspective, showing that urban forms developed in response to unique contemporary conditions. She notes, however, that grounding the city within the 'Arab-Islamic cultural domain' serves among other things 'processes of political legitimation' – a fascinating argument deserving of further study (Fuccaro, 2001). In spite of all this scholarship one is still confronted with contemporary observers continuing this 'Orientalist' trend. A recent example is by Barry Rubin, who, in an article titled 'Globalization and the Middle East', notes the following: 'Probably no area in the world resists – at least explicitly – globalization to an extent equalling that of the Islamic Middle East'(!) (Rubin, 2003). One is tempted to ignore such outlandishly exaggerated statements, yet they belie an emerging trend among many proclaimed 'Middle East scholars' and – more dangerously – policy-makers.

What is largely ignored, or left out, is that Arab cities have been part of modernization efforts from the nineteenth century to the present. While being subjected to colonialism, they nevertheless were able to grow, develop and contribute to architecture/urban design to an extent that has not been examined thoroughly in the literature. Urban forms unique to each city were developed, responding to larger 'global' issues. For instance theoretical constructs such as Ebenezer Howard's 'Garden City' found application in places as diverse as Cairo as pointed out by Khaled Adham in his chapter. Or, the rich architecture which flourished at the time of French colonialism, described by Bechir Kenzari in his chapter on Tunis. That trend continues and, within the current discourse on globalization, the Arab city has responded in a unique manner which illustrates that it is not just an 'Arabian nights' fantasy – but a real, vibrant, cosmopolitan entity which does not differ from anywhere else in the world – whether in the aspiration of its citizens or in their daily struggle to make a living. On the other hand, cities such as Sana'a have been in a state of underdevelopment due to their isolation from worldwide developments until the 1960s, as has been poignantly illustrated in Al-Sallal's chapter. That cities do or do not progress could thus be primarily comprehended by the extent to which they were subjected to colonialism – in a way allowing them to interact with the rest of the world. The exclusively 'Islamic' reading of such cities is – to put it bluntly – outdated and counterproductive. Heritage and culture are vital issues, but they should not be the sole, or dominant, factors through which the Middle Eastern city is studied and analysed.

The existing literature on Middle Eastern cities tends to amplify the trends noted above by having a very strong historical focus, such as AlSayyad's morphological analysis of Fatimid Cairo (AlSayyad, 1991), Basim Hakim's legal discourse – mentioned above – and its relation to the spatial morphology of Tunis (Hakim, 1986), and Jamel Akbar's analysis of the Arab-Moslem city (Akbar, 1988).[2] There are of course exceptions and some scholars employed a postcolonial framework, examining the impact of colonialism on 'traditional' societies, transformations due to national movements, and the construction of identities. These would take the form of case studies, looking at specific events, buildings and so on. An interesting contribution, falling within this category is a collection of essays exploring streets throughout the world; these include a number of Middle Eastern 'sites' – Cairo, Istanbul, Tripoli, and Rabat (Çelik *et al.*, 1994). In that same vein another collection of essays examined various 'forms of dominance' (AlSayyad, 1992). Other attempts were made by geographers, anthropologists, historians and political scientists to understand the spatial character of 'Arab/Moslem' cities (for example, Mitchell, 1988; Mostyn, 1989). And there is, of course, the classic work of Abu-Lughod on Cairo tracing its development from its conception until the 1960s (Abu-Lughod, 1971). Interestingly, Abu-Lughod was one of the first to criticize the 'monolithic Islamic city model', finding great similarities between this model and cities in medieval Europe and arguing that similar economic factors led to specific city forms (Abu-Lughod, 1993).

A more recent addition is by Stefano Bianca (2000) where an attempt is made to establish a religious basis for the urban form in the Arab world. It includes four case studies: Mecca, Baghdad, Fez, and Aleppo, and it is quite telling that more than half the book is devoted to an examination of the historic structure of so-called 'Arab-Islamic' cities, thus setting the stage for what is termed a conflict between 'traditional Islamic concepts' and 'modern Western planning principles'. The mere existence of such a model is never questioned – it is assumed that it is there and any subsequent examination of contemporary case studies is done in relation to this assertion, with all the conclusions that follow. Buildings and urban projects are thus evaluated and criticized based on the extent to which they deviate from, are an approximation of, and merge with, this historic model. The mere notion that the Arab (Moslem) could be capable of producing contemporary, modern structures, which do not necessarily have to be reflective of any illusive traditional concepts but are a unique response to contemporary conditions, is nowhere to be found.

Such a trend – establishing a religious undercurrent to the spatial structure of Arab-Moslem cities – is in fact a recurring theme. A recent review of the literature from a geographical perspective is littered with statements highlighting this. In one paragraph it is noted that in 'Islamic cities' there is an '. . . interconnection between the spiritual and political which, unlike in Western cities, is a major force in shaping the Islamic city' (Stewart, 2001, p. 177). Furthermore, the author notes, 'despite disagreement over the Islamic city model, it is undeniable that Islam, as a socio-religious system, has had a significant impact on the morphology of cities where it predominates' (p. 177). What is peculiar is this insistence on linking the religious and the secular, an attempt to ascribe to a religion some planning guidelines which were never intended to be there in the first place. Even if one admits, for argument's sake, that such a link may exist, it is not a very useful model to begin with. It leaves the reader with the impression that Moslems have been unable to go beyond the twelfth century, and that attempts at modernization will thus ultimately lead to the destruction of their heritage, identity etc. A much more useful model would be to examine contemporary social, political and economic factors operating at a global level and to understand the extent to which they have shaped the Middle Eastern city. The work of Çelik (1997) on Algiers's development and the extent to which it was guided by colonial policies could be an example. One may also examine the impact of capitalism, socialism, and the recent global economic paradigm and the free flow of goods, people and ideas – the focus of this book. In addition the whole notion of modernity and its perceived incompatibility with Islam, by being essen-tially a Western construct (Giddens, 1990), needs to be understood particularly in its spatial manifestations. For the Moslem to become 'modern' s/he must become, and this is always implied, Western – hence Western urban models should be accepted unquestionably. The transformation of the modernist project into the current globalization discourse evokes similar arguments.

While the literature is filled with work examining the colonial impact on the urban spatial structure of Middle Eastern cities, there is a gap when it comes to studying the influence of contemporary global trends – namely globalization. For many in the Middle East this represents a continuation of colonial policies, albeit in a new form not necessarily tied to physical domination. The American occupation of Baghdad which took place at the time of writing this book is a notable, and peculiar, exception. However, in general, there is a perception that other more subtle means are used to ascertain the hegemony of newly emerging world powers (the IMF, World Bank, multi-national corporations). While prevalent, such a view is largely based on emotional responses and should be examined from more than one perspective. In particular a pressing question needs to be answered: is the response of the Middle Eastern city to these globalizing influences an echo of similar trends in other cities? Is there evidence of a new spatial order due to globalization? I will try to discuss this in the following section.

Globalization, Cultural Hybridity and Polarization

Globalization has become, within the last few years, a catch phrase usually mentioned in a negative context and linked to the end of the twentieth century and the beginning of the twenty-first. In architecture and urban design it is typically associated with a loss of place, identity, and character. Cities throughout the world are becoming shapeless entities with a haunting sameness. It is in Third World metropolitan areas where this 'universalization' acquires greater urgency since it is associated with Western hegemony. Such a view has been eloquently articulated by Edward Said in his landmark *Orientalism*, which paved the way for what has been commonly known as 'colonial discourse analysis', where he argued that there is a 'culture of imperialism' in which the West is trying to superimpose its values and beliefs on the East (the focus of Said's analysis) (Said, 1979). Jane Jacobs noted that such an approach led to the creation of a new analytical language in which such constructs as hybridity, diaspora, creolization, transculturation, etc. figure prominently (Jacobs, 1996). Nezar AlSayyad extends this analysis further by arguing that in cities throughout the world globalization led to creation of 'third places', in-between spaces of 'spatial reconciliation of incommensurable constructions of subcultures' (AlSayyad, 2001). There seems to be a trend in the literature however that globalization does not in and of itself lead to loss of identity/heritage. Jacobs articulates this view as follows: '. . . globalization does not signal the erasure of difference but a reconstitution and revalidation of place, locality, difference' (Jacobs, 1996). This view is also echoed by AlSayyad who writes that '. . . the history of the world demonstrates a movement toward cultural differentiation and not homogenization' and '. . . urbanism will continue to be an arena where one can observe the specificity of local cultures and their attempt to mediate global domination' (AlSayyad, 2001) (figures 1.2 and 1.3).

Yet there are some common issues which are beginning to emerge, described as a 'new trend', which could be linked to globalization. In particular the relevancy of the nation-state is questioned

– in that respect city governments are emerging as the new centres of the 'new' global economy. Transactions occur between multi-national corporations, financial centres, and cities. However, it is noted that the nation-state still has a role to play, albeit through a restructuring process involving a move into supranational levels (the European Union or the Gulf Cooperative Council, for example). But cities are assuming a powerful role, and as a result of such processes they are increasingly being viewed as a product that needs to be marketed. These marketing efforts involve attracting headquarters or regional branches of international companies and staging of 'mega-events'. Other projects include luxury housing, dining establishments and entertainment amenities to attract the professional personnel required to operate these global activities. Urban projects, such as trade centres, conference centres and

Figure 1.2. Central Hong Kong: an intensification of locality within a global world.

Figure 1.3. An Egyptian restaurant in Central Hong Kong.

hotels, provide a catalyst in further encouraging investment and tourism. Architecture in many instances is used as a tool to create 'eye-catching' impressions – the Guggenheim museum in Bilbao is an example in that regard (Habitat, 2001). Such projects are the means to revitalize an otherwise 'stagnant' city – a process sometimes called the 'Bilbao effect'.[3]

As a result of all this some have noted that a 'dual-city' is emerging in which social polarization is becoming a dominant feature. The work of Saskia Sassen is perhaps representative of such a viewpoint (Sassen, 2001). Due to the presence of these high-profile projects there is an influx of a highly skilled, and paid, workforce. To maintain and service such activities, however, low-wage employees are needed who form the backbone of corporate and financial activities (figure 1.4). Furthermore, a 'rub-off' effect on local industries has been questioned by research. Thus, a geographical/spatial division occurs in which there are areas with a high concentration of poor in contrast to enclaves housing the very rich. Needless to say such disparities provoke '. . . resentment, social instability and conflict' (Habitat, 2001, p. 30). In other words these mega-projects do not necessarily lead to the happiness of the cities' residents since they are geared to a certain class.[4]

One of the most visible aspects of globalizing cities is what has been sometimes described as the 'quartering of urban space' due to a sharper division between rich and poor.[5] This quartering manifests itself in the presence of distinct residential cities – the most distinctive of these is the enclave or 'citadel'.[6] These are '. . . areas that can be considered as protected enclaves of the rich, the representatives of an extremely mobile top, operating at a more global level than ever before . . . [they] generally consist of expensive

Figure 1.4. Residents in Dubai using the low cost *abra* or ferry.

apartments in favorable locations' (Marcuse and Kempen, 2000, p. 4). Examples for these citadels are gated communities, private, high-rise condominiums with heavy security, to cite a few. In simple terms it is an effort to 'wall some in and keep others out' (Habitat, 2001, p. 30).

How does the Middle Eastern city relate to these developments? As the book will show it does not operate in a vacuum but relates to such issues in a strong manner. In some instances the response could be quite enthusiastic and exemplary, such as in Dubai, for example, while in others the attempt to join this 'international party' remains elusive – Algiers for instance. In general, however, the Middle Eastern city has strong potential for examining these issues for a variety of reasons. Many have been subjected to colonialism in the eighteenth and nineteenth centuries, which in many respects is very similar to contemporary conditions in terms of the political reality (Western hegemony) as well as economic conditions (capitalism). Strategies, at the urban and architectural level, were developed to cope with this context. For example, attempts were made to wipe out the past and join the 'civilized' West. Yet it is interesting to note that these changes at the turn of the twentieth century occurred within an overall climate of 'modernity' in which history was abandoned in favour of 'embracing' a new age. These developments had a significant impact on the spatial structure of many cities in the region – for example cutting wide thoroughfares through the dense urban fabric of historic quarters and superimposing spacious plazas, as has been pointed out by several contributors in this book, especially in 'traditional' centres – Algiers, Tunis, Cairo and Baghdad.

If this was the situation at that time could one draw parallels to contemporary conditions? Many would argue that there are strong similarities to the political climate present today. In that respect 'colonization' has been replaced by 'globalization'.

This is a tempting viewpoint since globalization, while being presented as a benign process which involves primarily the free flow of goods, people, and information, also contains an element of domination: whether political, social or cultural – since the relationship between developed/underdeveloped nations is not one of equality. But other similarities exist as well, in particular with regard to what one might term a paradigm shift. Similar to contemporary advances in information technology and global communications, technological and artistic advances at the beginning of the twentieth century revolutionized people's sense of space and movement which was reflected in a 'new architecture' and 'urbanism'. The questions thus remains: how has the Arab city responded to globalism and – this is perhaps the most contentious issue – does it differ significantly from colonialism?

The Structure of the Book

The book explores these issues of identity, globalization, hybridity, colonization and the struggles, strategies, and solutions offered by a selected set of Middle Eastern cities. The selection of these cities was based on a series of factors. My main concern was to include a variety of 'types'. These would include cities with a rich tradition, the ones one would immediately conjure up on hearing the phrase 'Middle Eastern' (Cairo, Tunis, and Baghdad); cities that have not been explored sufficiently, or 'forgotten' cities (Algiers and Sana'a), in addition to newly emerging Gulf 'oil' cities (Dubai). While any classification system tends to be problematic is was nevertheless a starting point to move away from that monolithic entity referred to as the 'Arab/Moslem' city.

Furthermore, authors were selected on the basis of belonging to, being raised in, or being a citizen of, the city. A major shortcoming of the literature is the lack of a 'local' voice and I was hoping through this strategy to project an empathy and a deeper understanding which may be lacking if the writer is an 'outsider' – which may however in some instances have its benefits allowing for an objective evaluation of events. All authors are young, Arab scholars educated in Western universities (US and UK) and their views provide a fresh, and energetic, view of their cities. Another factor relating the authors together is that they are all architects, hence they have a natural inclination towards focusing on the built environment, i.e. emphasizing the spatial aspect. As such the chapters are, on the whole, heavily illustrated. These images are seen as critical in illuminating the themes explored in the book. It must be noted that in spite of many commonalities among the authors they prescribe to different ideological and theoretical viewpoints to foster the multi-representational framework discussed above.

Following this selection process a series of questions was posed to the contributors:

1. To what extent did colonization influence the shape, urban form, and architecture of the city?

2. What mechanisms were developed to protect heritage and identity?

3. Using the idea of hybridity, how did respective cultures incorporate or integrate with foreign influences?

4. Does the contemporary political climate of globalization, which dominates the urban and architectural discourse, have any spatial influence or manifestation?

These questions were examined by the contributors. Freedom was given in terms of scope and scale of contributions – with the exception that the time-frame to be covered should be from late nineteenth century to the present day. The chapters were initially received in draft and the drafts were circulated among all authors with comments pertaining to each chapter. Two divergent approaches emerged in response to these preliminary contributions and ensuing discussions. First, one group provided a general overview of their cities discussing colonial policies and the impact of colonization in general (Algiers, Baghdad, Sana'a); the second group – while providing a general overview as well – did, however, examine specific 'case studies'; these include Cairo (the district of Heliopolis and Dreamland), Tunis (the Lake Tunis project) and Dubai (Burj Al-Arab Hotel and the Palm Islands). Thus the book has been divided into two sections: (1) Overview and Developments and (2) Case Studies: Redrawing Boundaries.

The book begins with what is in effect a prologue, written by well known author and expert on Middle Eastern cities, Jamel Akbar, titled 'Merits of Cities Locations'. He offers us a daring, and some might say 'utopian', view with respect to globalization in an Middle Eastern context – essentially adapting the polarization argument discussed above and offering an alternative view with regard to the location of cities. Cities, according to Akbar, should be located next to 'sites of merits', i.e. resources; in that way they will cease to be 'parasitic' entities. I will return to his essay in the final part of this introduction.

Overview and Developments

This section contains three cities: *Algiers*, *Baghdad*, and *Sana'a*. Their authors have elected to provide a general overview of their selected cities prompted in part by a realization that they have not, for a variety of reasons, responded to the globalizing influence occurring throughout the world. There are several reasons for this: terrorism in Algiers, and the Gulf War in both Iraq and Yemen, which have all been external factors hindering the emergence of these cities as global players. None of these cities has engaged in the kind of large-scale urban development witnessed elsewhere in the region. One city in particular – Baghdad – is at the time of writing this introduction still under occupation, thus bearing the full brunt of a global, hegemonic force, which in turn has put its development on hold.

The first contribution in this section, by Karim Hadjri and Mohamed Osmani, is '*The Spatial Development and Urban Transformation of Colonial and Postcolonial Algiers*'. Their main focus is to examine the urban transformation of Algiers during and after colonization from 1832 until the present. They show that the initial French intervention was founded on a double conviction: the superiority of the Western civilization and the inferiority of the indigenous civilization which in turn was reflected on spatial policies. The '*Casbah*' – Algiers old city built by the Turks – became a setting for the display of colonial policies. Hadjri and Osmani point out that the French did not originally consider the possibility of creating a new urban structure next to the pre-existing urban fabric, which was typical of French colonization in other countries. The destruction of various parts of the Casbah reflected the military character of the intervention.

The authors note that the French did, however, eventually create a pre-colonial city named *Place d'Armes* to allow the gathering and parades of troops. Many streets were straightened and widened and various buildings were transformed and adapted for new needs. It should be pointed out here that this situation was not unique to Algiers; in fact such undertakings were quite common in Europe – particularly in Paris, but also in Rome as pointed out by Spiro Kostoff (Kostoff, 1994). These developments, however, were marked by an incoherent policy leading to a continual conflict between the concept of Algiers as a military centre and the idea of developing the city into a commercial emporium. Hadjri and Osmani note that the transformation of the conquest into colonization attracted new immigrants who settled in the 'Westernized' part of the Casbah, almost depopulated of its Moslem citizens. Among the urban manifestations of this was the arcaded Boulevard de la République which, through its monumental façades, became a symbol of colonial Algiers for the people coming by sea from France, reassuring them of being home. They point out, however, that 'it was only an architectural screen hiding uncontrolled development behind'. Some of these interventions were almost exact copies of similar projects in Paris.

The authors further describe the urban development of Algiers between 1880 and 1914. One of the main factors contributing to urban and spatial growth was the success of viticulture in Algeria which generated considerable wealth. Mainly geared towards French consumption, funds became abundant giving an impulse to the process of urbanization in the whole colony. Hadjri and Osmani illustrate, however, that this had an impact on the 'traditional' centre. The Casbah was emptied of its original functions and then of its position as an urban centre, becoming a 'marginal' quarter, 'dilapidated' and crowded with an indigenous proletarian population. The 1930s saw the emergence of shanty towns or *bidonvilles*. High population growth created a drastic housing shortage prompting migrants from the countryside, who could not find living spaces inside the Casbah, to construct their own shelters on unbuilt areas. These spontaneous settlements, which did not follow either the Casbah's pattern or the European model, formed the *bidonvilles*, which gradually ringed the city. The authors describe Le Corbusier's visit during that time and his (failed) attempt to impose his urban theories, developed in France, on Algiers – also termed 'the new bombardment of Algiers'. The authors go on to discuss the influx of foreign architects in the 1970s (the socialist period) and their impacts on the city's urban structure, in particular the work of Oscar Niemeyer who was involved in building many large-scale projects. They note that this period saw greater international input than in the 1980s and 1990s when 'internationalization passed the city by'.

Finally Hadjri and Osmani argue that globalization did not affect Algiers in the way it did in other African and Arab cities. They note that Algiers has retained its 'corner shop' culture and did not open fully its doors to multi-nationals and the movement of capital. It is implied here that local politics have played a role in the city's isolation and in turn its inability to interact globally. In addition, tourism never really blossomed either in Algiers or the rest of Algeria. This has led to the decline of the Casbah, effectively becoming a ghetto. As a result Algiers continues to grow as a group of large independent neighbourhoods rather than as a global entity. They posit a set of measures, which could help in the proper urban development of the city, mostly centred around the creation of an investment climate to attract businesses, tourism etc., thus spurring urban growth. The authors here are in some ways looking further to the 'East' – Tunis, Cairo and, perhaps, Dubai – searching for an appropriate model for their city.

The following chapter by Hoshiar Nooraddin, is titled *'Globalization and the Search for Modern Local Architecture: Learning from Baghdad'*. Nooraddin argues that events from the historical development of Baghdad's architecture during the twentieth century show an inability to interact with international architectural movements. This contributed to dramatic consequences on both the local architecture and the city's transformation. His chapter discusses the impact of different 'forms of globalization' during the twentieth century on the contemporary architectural and urban design character of Baghdad. In doing that Nooraddin explains the role of international architectural firms, local architects, and local users in this process. Furthermore, he shows how the shifting political and economic systems of the dominating ruling power influenced this process and how these shifts have affected the urban development and architectural character of Baghdad. These issues are examined through a historical perspective to provide 'lessons to improve the process of developing local architecture in the present globalization era and the emergence of a "new" Iraq following the power shift in 2003'. These lessons could also be important, argues Nooraddin, for the post-war reconstruction projects in Iraq. His chapter was written at the time of the US led war on Iraq and thus had to be constantly modified based on changing political realities, yet it provides some interesting insights into the impact of international sanctions, and the perpetual state of war on the city, its architecture, and its citizens.

Nooraddin argues that, since the colonial period in the 1920s, the Iraqi construction market increasingly attracted large numbers of international architectural and construction firms. They played a significant role in shaping the contemporary character of Baghdad's architecture. The outcome of this process is a city with a variety of architectural styles. Furthermore, the political instability and power shifts during the twentieth century contributed to 'diffuse' Baghdad's contemporary architectural character, such that each area of the city developed its own character and the only real relationship between areas is the street network. In one interesting passage he notes that in the 1950s the government removed traditional balconies along Al-Rasheed Street to allow enough space for imported London buses, an episode which was symptomatic of the large-scale changes occurring within the city. He notes that at that time there was an increase in squatter settlements.

Nooraddin poignantly describes the role of local architects and their attempts to create an architecture shaped by local needs, while increasingly hindered by political ideologies whether related to colonialism or the subsequent emergence of the Ba'athist regime. He presents and critiques the work of two notable architects, Chadirji and Makiyya, placing them within global architectural trends and their struggle to establish a 'local' architectural language. He reveals, however, that in many respects their attempt to 'import' Islamic design principles from cities throughout the region does not differ fundamentally from being influenced by Western planning and architectural principles – neither necessarily relates to the needs of the local population. It is worthwhile noting here that many Arab architects in other countries were more or less engaged in the same endeavour. Hassan Fathy in Egypt, or Rasem Badran in Jordan just to cite a few were all searching for ways to articulate an indigenous architecture, derived from 'cultural roots'. These developments occurred within an overall 'global' climate which saw a dissatisfaction with modernism and an attempt to 're-connect' with history.

The author concludes with a set of directives to help in the current reconstruction effort. Primarily, he notes that the involvement of international firms – due to the 'globalization' culture – should be balanced. This balance could be achieved through a set of guidelines which include: the development of local building regulations; the development of local architectural knowledge; and an interactive involvement of local architects in that process.

The last chapter in this section by Khaled Al-Sallal is titled ' *Sana'a: Transformation of the Old City and the Impacts of the Modern Era'*. A 'traditional' reading of a fascinating city which, unlike any other city in the Arab world, acquires its identity through its unique 'vertical' architecture. Unprecedented in the world, it is only natural for any author attempting to find a place for Sana'a in the contemporary globalization discourse to deal with this traditional setting. Al-Sallal, offers an environmental perspective, in which he uses a socio-spatial model to extract principles which led to the success of Sana'a's traditional environment. He also points out that unlike many cities in the region Sana'a was isolated from the world beginning in 1919 – during the emergence of the 'Imam' regime. At a time when modernity began to influence the Arab world through colonization, Sana'a remained in isolation – which in turn inhibited development – until the 1962 revolution which opened it up to the rest of the world.

The chapter starts by highlighting the issue of architectural transformation in Sana'a after the Yemeni revolution in 1962 – a significant turning point in its modern history – and the factors leading to this event. A rupture occurred between the traditional vertical tower house and the horizontal Western style villas which began to dominate the cityscape. The new transformed architecture has – according to Al-Sallal – no identity, lacks order and unity, is alienated in its society and environment, and gives minimum consideration to comfort issues. The chapter then raises an important question 'Why has vernacular architecture in places like Sana'a proved to be more successful than contemporary architecture?'. To answer this question, the author argues that there is a need to understand the intricate relationship between culture, environment, and architecture in the traditional society of Sana'a. He notes that the literature has not gone beyond the level of historical and physical description and has not investigated other factors – for example, environmental ones.

Al-Sallal describes the historical development of the city, starting from the pre-Islamic period until the Islamic period when significant transformations took place. Following this historical background, he explores the tremendous changes affecting Yemeni society after the revolution and how these changes influenced the city and its architecture. This transformation is discussed from the social, economic, and environmental points of views. The author argues that the vast growth of the city population, the migration from the villages to Sana'a, and the return of labour migrants residing in the Gulf were among the major factors that caused these changes. The chapter further explains the vernacular architecture of Sana'a from an urban and architectural perspective; this is followed by an analysis of the architecture that emerged after the revolution in the new housing districts of Sana'a – viewed by the author as a 'drawback', in relation to the traditional settings. He makes a comparison between the traditional and the new architecture with regard to urban pattern and spaces, building forms, façade and window design, interior space design, and building construction.

The chapter concludes by drawing attention to the importance of preserving the heritage of Sana'a, which is in danger of having its position as a world heritage site removed by UNESCO. The author recognizes that social changes preclude people from living within the traditional environment, in fact he writes that the city has 'two disconnected worlds' – the traditional and the modern, thus creating a sort of 'rupture' for its citizens – either live in the past or become modern. He nevertheless points out the importance of using the traditional setting as a resource. In that way it may offer 'cultural and ecological' values to create an appropriate contemporary urban environment.

Case Studies: Redrawing Boundaries

All three cities in this section have made attempts to become global players, with varying degrees of success. A central trait, shared by all three, is that they have engaged in large-scale urban projects meant to attract visitors and investors to the city – in part encouraged to do so by being shielded from political turmoil. These projects are characterized by their exclusiveness and inaccessibility – a 'redrawing of boundaries'.[7] Whether it is the creation of a new city centre (Tunis), the proliferation of gated communities (Cairo), or the emergence of palm-shaped islands (Dubai), these are all an attempt to respond to, and engage with, the new global economy. Unlike cities in the first section, which to some degree have not been able successfully to overcome their colonial past, one is tempted to attach to these cities the cliché 'success story'. Yet that may be a premature judgement, as the authors describe in great detail. Aside from the divergent historical circumstances between Dubai on the one hand and Tunis and Cairo on the other, economic circumstances have also contributed to the emergence of Dubai as a truly global centre, a 'crossroad of the new global economy'[8] – which has led some to describe it as the most globalized of all Arab societies. Given the perceived dominance of Dubai, a shift has occurred whereby influence is exerted on other traditional cities such as Cairo, which is emulating Dubai in many of its endeavours (Summer Shopping Festival, for example) but also in large-scale investment by Dubai based businesses. Thus, examining all three cities is viewed as critical to the main arguments which I have presented above – and offers a counterbalance to the previous section. Furthermore, they can provide potentially useful lessons for other cities in the region.

The first chapter, *'Lake Tunis, or the Concept of the Third Centre'*, is by Bechir Kenzari. He offers a unique, and fascinating, reading of a Middle Eastern city and its relation to globalization through an ecological perspective. He identifies within the city of Tunis 'three forms of centrality'. The traditional *medina*, the European city, suburbs built in between and – the main focus of his chapter – a third centre, the Lake Tunis projects (Lac Nord and Lac Sud). He notes that this 'city within the city' is not directly related to other parts of Tunis, that it caters to hi-tech and international clients, effectively becoming a centre for diplomacy and international business. Further enhancing its 'international' appeal, it is located next to the airport. Prior to discussing these projects however, Kenzari provides a historical framework.

The author begins by describing the traditional *medina*, its Ottoman origins and the extent to which it suffered from colonialism, which began with the French in 1881. Similar to developments elsewhere the French surrounded the *medina* with a ringroad, and a European city housing non-Tunisians encircled the *medina*. Kenzari describes this as follows: 'The old city found itself strangled from all sides by the symbols of French might, the menacing signs of power'. As a result at the time of Independence in 1956, the *medina* became a 'slummy, decaying' area. The houses were for the most part occupied by migrants from the countryside. Kenzari writes that due to this deterioration the government tried to impose 'slum clearance', a 'modernist vision of spatial organization' which led to the destruction of large areas within the historic fabric to facilitate vehicular traffic.[9] In examining the new European city – the 'second form of centrality', the author notes that it was laid out according to Western planning principles, which was a typical and expected intervention of colonial powers. However, Kenzari offers a picture of a rich and diverse architecture which flourished at that time. An architectural style incorporating 'Tunisianity' was used by the French, not so much to become symbols of France but – according to Kenzari – architectural portraits of Tunisia – thus painting an unusually benign picture of colonial hegemony. Subsequent developments saw the emergence of suburbs which are classified according to different criteria, the most dominant of which is the emergence of uncontrolled, spontaneous settlements.

All of the above forms a backdrop for Kenzari to discuss in detail the Lake Tunis project. Conceived as an ecological endeavour, the author delves in great detail in describing the various environmental problems which led to the decision to sanitize the lake – initially a source of pollution – and reclaim land. He notes that through this project a new city is emerging, facing the Mediterranean (Europe?) and 'somehow turning its back to the old medina and European city'. The project began in the mid-1980s in an effort to address the complex geographical problems of Tunis, with Middle Eastern investment and Dutch technology. In the process hundreds of hectares were reclaimed and a new urban zone emerged. The project is viewed by Kenzari as a direct response to globalization – in addition to the ecological factor. In particular he points out Tunis's emergence as a leading site for financial and specialized services, enhanced by the economic association agreement signed with the European Union in 1995. Such a 'European' orientation – emphasized by the author – should not, however, detract from the possibility of other influences. In fact parallels could be drawn with Dubai, particularly with regard to land reclamation efforts occurring there as well (the Palm Islands).

Kenzari concludes by noting that the Lake Tunis Project emerged 'as a whole and out of a bi-polar situation (*medina* versus European city)', and he reminds us that such transformations illustrate 'the macro-geographic disequilibrium of the city and its history, but also the contradictions between the goals of planning and the weight of market and political decisions in the age of globalization and ecological consciousness'.

The following chapter by Khaled Adham, is titled *'Cairo's Urban Déjà Vu: Globalization and Urban Fantasies'*. Adham takes us in his narrative on Cairo on a journey spanning a century – from early twentieth century to the present. Wonderfully written, it deals with some of the most critical issues in this book by juxtaposing two projects, two developers, and two time-zones. By this means he hopes to show that capitalism/colonialism and globalization are, more or less, two sides of the same coin. While the focus is on Cairo, he reads its contemporary transformation through two real estate ventures, each representing their age and responding to various economic circumstances. He begins by asserting that 'many of the circumstances that ushered the economic upsurge of the 1890s resembled those that led to the upswing of the 1990s', in particular with regard to real estate development. He attempts to 'provide through a back and forth excursion in historical space and time . . . an overall view of the influence that international capitalism has exerted upon Cairo in the opening and closing decades of the twentieth century'. Furthermore, the integration of Egypt into world culture '. . . influenced the production of two selected land developments, urban fantasies, from these two eras, namely, Heliopolis and Dreamland'.

Each of these developments is discussed within its respective socio-political and economic backgrounds. Thus, Adham begins his reading of Cairo by discussing what he terms 'the colonial economy and agrarian capitalism'. In it he deals with such issues as the history of private property, and the emergence of a modern market system. One event naturally stands out as pivotal in the emergence of modern Cairo – the opening of the Suez Canal in 1869. This is discussed in great detail particularly in relation to '. . . channelling a vast flow of international traffic through the country [and] increasing the linkage with the outside world': the first step towards drawing Egypt into the 'system of expanding Western capital economy'. Adham follows this with a brief discussion of the urban and architectural development witnessed within Cairo at that time and its continuous expansion, particular with regard to residential developments. Within that background he discusses the origins and development of Heliopolis, which is what he terms 'a European vision of the Orient'. He shows us how this particular development was linked to 'the global forces of its time' and – most importantly – how it 'reflected this growing fantasy with the Orient'. Adham argues that the district was based on the 'Garden City' concept as developed by Ebenezer Howard – although not completely faithful to its utopian ideals. He notes that the founder of Heliopolis, the Belgian Baron Empain, eliminated the socialist agenda from the concept of the Garden City and re-invented it solely in formalistic and marketable terms'. By using neo-Arabic elements in the architecture of this new city Empain wanted to create a tourist destination – in short a fantasy environment.

Adham then jumps to the late twentieth century, in particular the year 1991 which in his view is the starting point of the 'postcolonial economy and mixed capitalism', due to various political factors occurring at that time resulting in Egypt's debt forgiveness and in turn leading to a change in the city's urban and spatial structure. These economic events are discussed in detail and it is interesting to observe, as Adham points out, the role played by such organizations as the IMF and the World Bank which 'imposed' on Egypt what has come to be known as SAP – Structural Adjustment Program. According to the author these organizations '. . . became the principal tools through which the ideology of neoliberalism has been promoted and policed'. Such policies have led to the 'commodification' and 'privatization' of spaces for 'cultural consumption'. One of the real estate ventures emerging against this backdrop is Dreamland which, Adham writes, is an 'Oriental Vision of the Occident' – an interesting reversal of words. Essentially a gated community for upper-class Egyptians, it is designed to cater to 'Western tastes' evidenced in the choice of architects/developers to realize this vision – mainly North American developers who are also engaged in similar projects throughout the Arab world. The author notes that there is a crucial difference in the approach towards the design of Heliopolis and Dreamland: whereas the former was based on the 'Garden City' ideal, the latter is more or less using 'Disneyland' as its inspiration – a place 'designed solely for consumption and fun'. This is evident in the names given to various residential buildings – more or less perpetuating 'mythical' environments as one might find in any real estate venture anywhere in the world.

Adham concludes by noting that projects such as Dreamland are an attempt by Egypt to join 'the club of world cities'. On the notion of authenticity with regard to these two developments he writes that 'Heliopolis was [not] any more authentic than the Dreamland of today. Both developments depended on the production of fantasy and entertainment for promoting their commodity'. In fact he argues that they are 'deliberately formulaic and fabricated illusions of the Orient and the Occident'. Global influences, whether at the turn of century or today have '. . . always been enticing the desires, which in turn stokes the construction of urban fantasies such as Heliopolis and Dreamland'. According to Adham this is why for many historical observers of Cairo it seems that they are '. . . having a *déjà vu* experience'.

I am the author of the last chapter in this section, *'Redrawing Boundaries; Dubai, an emerging global city'*. It is perhaps appropriate to conclude this journey in the Arab world with a newly emerging city, which only began to be considered a significant urban entity in the late 1960s. At the present time it is a source of envy and emulation among its regional counterparts. Its architecture, events and megaprojects have become a basis of discussion, research and – in many instances – pure puzzlement.[10] However, here I expose some of the problems which are associated with such rapid urban growth, by examining the region's history and the transformation witnessed within the twentieth century. The Gulf region in particular has been an area of contention and struggle since the Portuguese attempted to control the vital spice trade in the fifteenth century. They were followed by the British whose policies led to the emergence of modern Gulf States including Dubai. More recently American hegemony is asserting itself in the region. Therefore, more than any other Arab city Dubai lends itself to being a 'global' entity and could be a potential model for other cities in the region.

In this chapter I focus on two projects which have come to symbolize Dubai's emergence as a global entity: Burj Al-Arab, a 7-star hotel shaped like a sail, and two palm shaped islands (Palm Jumeirah and Jebel Ali). The hotel's profile has been placed on car licence plates and the islands have become a heavily advertised icon. In critically assessing these projects I place them first within the current globalization discourse with its emphasis on the notions of hybridity and the inevitable outcome of 'polarization'. I then describe the historical background of the development of the city with a specific focus on the rapid growth witnessed within the last two decades, illustrating the dramatic urban transformations within this century while highlighting its multicultural nature. This review is followed by a 'survey' of the current 'megaprojects' undertaken by the city. Dubbed by some as a 'city of many cities'[11] these projects are symptomatic of the transformation witnessed within the last century. I then turn to the two case studies, and attempt to show how they are a continuation of modernization efforts in Dubai which began early in the twentieth century, and that they represent a unique response to the current globalization discourse.

I devote some time to the choice of direct symbols such as the sail or the palm. I argue that these symbols reflect a desire to connect with the past and to construct an identity *vis à vis* a predominantly migrant population. Therefore, the Palm Islands or Burj Al-Arab can be understood or comprehended as a continuation of such thinking, albeit on a gigantic, i.e. global, scale. I further note that these projects are a 'perpetuation of mythical environments' illustrating how this is achieved in both. An interesting observation is made with regard to the Arabs 'fascination with the sea'. Moving away from the desert and building in the sea itself could thus be construed as a 'sign of modernity'.

My analysis of these projects is that while being admirable in their sheer scope and ambition they tend to highlight the downside of globalization. I argue that they 'emphasize "polarization" and "exclusiveness"'. Furthermore I point out that the unique situation of Dubai 'unburdened by history and heritage, led to a tabula rasa approach whereby the city is constantly being re-invented'. The city has thus become a model example for what is a typical development of global cities in the world, specifically in its 'constant redrawing of boundaries' and the 'evacuation of the national'. My main concern is with the fragmentation of the urban environment; areas of the city are associated with elite elements, geared towards tourists, real estate speculators and multi-national corporate executives, thus excluding 'regular folk'. I conclude by issuing a warning to officials: in their drive to make their city a global centre, they should uncover 'the real Dubai' which does exist along the shores of the creek and in its marketplaces. Heightened speculation and a focus on financial development will simply exacerbate negative trends found elsewhere.

Concluding Thoughts

The prevalent view on globalization is that it is a phenomenon externally imposed upon nations, which are helplessly witnessing their cities and countries taken over by multinational corporations. I would like to take the position of Saskia Sassen, however, who quite rightly argues that '. . . globalization is not simply something that is exogeneous' (Sassen, 2001, p. 347). She notes that while some processes may seem to come from outside – multinational corporations setting up regional headquarters, for example – yet these are processes which are activated from the inside by local actors. In her words the global city '. . . represents the endogenizing of key dynamics and conditionalities of the global economy' (p. 347) – view echoed elsewhere in the literature as I have pointed out above (for example, AlSayyad, 2001; Jacobs, 1996). The chapters in this book tend to confirm this position. This is particularly so in the case of the last three cities, Tunis, Cairo and Dubai which have made attempts – with varying degrees of success – at internationalization. Many of the projects discussed were initiated by local governments without any (overt) external intervention; in some, such as is the case in Tunis, they are meant to serve the local population and as such are quite successful. In Dubai they have become national symbols. Influences do occur however and the shift of power in the Middle East to the 'East' – i.e. the Gulf Region – and in turn a re-configuration of traditional cities to – for example – 'become like Dubai' are real and observable phenomena (figure 1.5). In Cairo and elsewhere (for example, Beirut) projects are developed with Gulf investment. Some may argue that this could be another form of 'local' colonialism.

While I would not necessarily conceptualize this in these terms, the whole notion of globalization being another form of domination (i.e. colonization) is an outdated one and has contributed to the underdevelopment of many cities in the region – and we have seen some examples in this book. It is not a benign process either and as such its shortcomings should be addressed. Jamel Akbar in his essay alludes to some of the problems associated with these globalizing trends – namely concentration of wealth – by examining Islamic traditions, which are a vital part of this region. Without falling into the trap of offering prescriptive solutions – a problem that plagues many Middle Eastern studies – his argument of sharing wealth, equal distribution of resources, and the location of settlements offers an alternative, or to be more realistic a modification, to what is now called globalization. In that way he echoes perhaps the quote with which I began this Introduction where the great German writer Bertolt Brecht noted that a city comes into being only through 'the wisdom of its citizens' – *all* its citizens and not just a select few. Akbar ascribes to an 'Islamic' perspective, i.e. deriving inspiration from the great Islamic tradition which may stand in conflict with some of the themes and arguments which I have presented at the beginning of this chapter. Yet, it is important to

Figure 1.5. An advertisement in Cairo for 'Maadi City Center' a Dubai based conglomerate specializing in mega malls. The advertisement notes: 'Enjoy the difference'. In the background appear incomplete housing units typical of poor areas in Cairo.

note – as Anthony King has so forcefully argued – our modern representation of the city should celebrate multi-dimensionality, and we should '. . . encourage parallel and conflicting representations to coexist' (King, 1996, p. 17). The sacred and the secular can coexist, and the perceived conflict between them should only help to enrich urban studies in our region.

Finally, this book has attempted to shift the 'globalization versus fundamentalism' discourse – in which globalizing tendencies are perceived as a threat to local culture – a popular view in our region. Instead, an alternative is proposed in which political and social underpinnings of urban design are dismantled, thus uncovering the potentially 'positive' qualities of 'globalization' which is moulded by city residents' needs and desires. In such a discourse globalization becomes an expression of the political climate in each country, modified by local conditions, and should be understood in that context.

NOTES

1. Middle Eastern is meant here to include cities of North Africa as well – culturally and socially they are associated with Middle East region.

2. It has to be noted that in the case of AlSayyad, for example, his work is purely urban history, in many ways an excavation of the historical origins of the city of Cairo and thus has relevance from a strictly historical and theoretical perspective. No attempt at projection to the contemporary situation is, or should be, made. In another publication, deviating from this historic perspective, he makes an interesting contribution by arguing that the Islamic city should be termed the 'colonial Islamic city' since Moslems themselves engaged in forms of colonialism in newly conquered cities (AlSayyad, 1992).

3. Other examples exist such as Liebeskind's, Imperial War Museum North in Manchester. Such trends began perhaps in the 1960s with the Sidney Opera House, which became a symbol for the city.

4. A United Nations for Human Settlement Report notes the following: 'The kind of projects described may generate alienation, as many local people find little for themselves in luxury shopping centres. Indeed, many of the new projects may worsen the city residents' quality of life through increased noise or congestion, or the loss of the opportunity to use the limited supply of key urban sites for other uses. Scarce public resources may be allocated in support of mega-projects intended to project a positive city image but with little value in the residents' daily lives. In many cases, the projects create islands of activity, oriented to those of means, surrounded by areas of poverty. This phenomenon has been called the "bubble effect"' (Habitat, 2001, pp. 30–31).

5. It could be quite rightly argued that cities always contained areas of rich and poor. But one may note that the speed of these developments, their intensity and prevalence could be quite different from anything that may have occurred in the past.

6. The term citadel was initially coined by Friedman and Wolff (1982). Citadels are enclosed, protected, insulated areas of upper-income residence, often, particularly if located downtown, combined with office and commercial uses.

7. A term introduced by Saskia Sassen in describing the physical changes occurring in global cities (Sassen, 2001).

8. The official logo by *Dubai 2003*, the organizing committee for the IMF/World Bank Meetings held in Dubai in September 2003.

9. Such developments were occurring in places as diverse as Paris, Rome and Cairo. For a more detailed review of this see Elsheshtawy (2000).

10. A recent conference at the Harvard Design School was devoted to a discussion on the emergence of Dubai as a global entity. Organized by the Department of Planning at Harvard, Dubai Municipality and Dubai Executive Office, it was intended as a showcase for the city. Experts from around the world were invited to try to explain this urban phenomenon. Interestingly, the conference did not degenerate into a propaganda tool but posed significant substantive issues related to the fast urban growth witnessed in the city. (*Dubai Conglomerated: the urban planning and design challenges to building a city of many cities*. Harvard Design School, April 18–19, 2003). Also see the following newspaper article: 'Dubai described as model city: participants in Harvard Conference laud Emirates cosmopolitan culture; several papers presented'. *Gulf News*, April 20, 2003, p. 6.

11. Hashim Sarkis, Harvard Design School, in introducing the conference on Dubai.

REFERENCES

Abu-Lughod, J. (1971) *Cairo: 1001 years of the city victorious*. Princeton, NJ: Princeton University Press.

Abu-Lughod, J. (1993) The Islamic city: historic myth, Islamic essence and contemporary relevance, in Amirahmadi, H. and El-Shakhs, S. (eds.) *Urban Development in the Muslim World*. New Brunswick, NJ: Center for Urban Policy Research, pp. 11–36.

Akbar, J. (1988) *Crisis in the Built Environment: the case of the Muslim city*. Singapore: Mimar.

Al-Hathloul, S. (1996) *The Arab-Muslim City: Tradition, Continuity and Change in the Physical Environment*. Riyadh: Dar Al-Sahan.

AlSayyad, N. (1991) *Cities and caliphs: on the genesis of Arab Muslim urbanism*. New York: Greenwood Press.

AlSayyad, N. (ed.) (1992) *Forms of Dominance: on the architecture and urbanism of the colonial enterprise*. Aldershot, UK: Avebury.

AlSayyad, N. (ed.) (2001) *Hybrid Urbanism: On the Identity Discourse and the Built Environment*. Westport, CT: Praeger.

Bianca, S. (2000) *Urban Form in the Arab World*. London: Thames & Hudson.

Çelik, Z. (1992) *Displaying the Orient: Architecture of Islam at nineteenth-century worlds fairs*. Berkeley, CA: California University Press.

Çelik, Z. (1997) *Urban Forms and Colonial Confrontations: Algiers under French rule*. Berkeley, CA: California University Press.

Çelik, Z., Favro, D. and Ingersoll, R. (eds.) (1994) *Streets: critical perspectives on public space*. Berkeley: California University Press.

Donald, J. (1992) Metropolis: the city as text, in Bocock, R and Thompson, K. (eds.) *Social and Cultural Forms of Modernity*. Cambridge: Polity Press.

Elsheshtawy, Y. (2000) 19th century globalization. *Proceedings of IASTE Conference 'The End of Tradition?', Trani, Italy 2000*. Berkeley, CA: University of California, Berkeley.

Friedman, J. and Wolff, G. (1982) World city formation: an agenda for research and action. *International Journal of Urban and Regional Research*, **6**(3), pp. 309–344.

Fuccaro, N. (2001) Visions of the city: Urban studies on the Gulf. *Bulletin of the Middle East Studies Association of North America*, **35**(2), pp. 175–188.

Giddens, A. (1990). *The Consequences of Modernity*. Stanford University Press: Stanford.

Habitat (2001) *Cities in a Globalizing World: global report on human settlements*. Nairobi: United Nations Centre for Human Settlements (Habitat).

Hakim, B. (1986) *Arabic-Islamic Cities: building and planning principles*. London: KPI Limited (reprinted by Columbia University Press, 2001).

Hamadeh, S. (1992) Creating the traditional city: a French project, in AlSayyad, N. (ed.) *Forms of Dominance: on the architecture and urbanism of the colonial enterprise*. Aldershot: Avebury, pp. 241–259.

Jacobs, J. (1996) *Edge of Empire: Postcolonialism and the City*. Routledge: London.

King, A. (ed.) (1996) *Re-presenting the City: ethnicity, capital and culture in the twenty-first century metropolis*. London: Macmillan Press.

Kostoff, S. (1994) His Majesty the pick: the aesthetic of demolition, in Z. Çelik, Z., Favro, D. and R. Ingersoll, R. (eds.) *Streets; critical perspectives on public space*. Berkeley: California University Press, pp. 9–22.

Marcuse, P. and van Kempen, R. (2000) *Globalizing Cities: A new spatial order?* Oxford: Blackwell.

Mitchell, T. (1988) *Colonizing Egypt*. Cambridge: Cambridge University Press.

Mostyn, T. (1989) *Egypt's Belle Epoque: Cairo: 1869–1952*. New York: Quartet Books.

Rubin, B. (2003) Globalization and the Middle East. *YaleGlobal Online*, January 16, 2003.

Said, E. (1979) *Orientalism*. New York: Knopf.

Sassen, S. (2001) *The Global City*, 2nd ed. Princeton, NJ: Princeton University Press.

Serageldin, I. and El-Sadek, S. (1982) *The Arab City: Its Character and Islamic Cultural Heritage*. Riyadh.

Stewart, D. (2001) Middle East urban studies: identity and meaning. *Urban Geography*, **22**(2), pp. 175–181.

Chapter 2

The Merits of Cities' Locations

Jamel Akbar

It is widely accepted, although not scientifically or logically proven, that the technical advancement of today's societies is due to capitalism. It is also widely accepted that globalization is the fruit of both capitalism and technical advancement (mass transportation, telecommunications, etc.). Since modern human societies have not had the chance to experience societal systems other than capitalism or socialism, accumulation of human knowledge has been largely associated with capitalism. It is not the place here to explain that the acceleration of human knowledge is not due to capitalism, nor is it also the place to argue that other societal systems if given the chance could produce more appropriate human knowledge than capitalism has produced (nuclear reactors for example). Yet, accepting the association between current human technical achievements and capitalism (although refutable), one should ask a central question: is capitalism as a producer of contemporary technical advancement the best among other systems to maintain and even develop such advancement? Or, could capitalism be the best societal system to continue governing today's global character?

If globalization has defeated both geographical and political borders, it is however characterized by its capacity to allow capitalists (corporations and the very rich) to seize all possible investments. Thus, as is pointed out by many scholars, globalization is unfortunately stratifying humans into two distinct groups, the rich and the poor. This is one major plight among many produced by capitalism. However, even if one accepts that other plights, such as pollution and environmental degradation, can one day be controlled under the existing internationally accepted democratic political structure (although refutable), then this plight arises as a serious one. Living those conditions would result in inefficient, crowded, and unjust cities. The nature, the morphological tissue, and the socioeconomic structure of our contemporary cities in the Arab Moslem world reflect this situation. These cities are inhabited by technocratic, consumerist individuals who have the right to empower some over others, and by the wealthy who have the capital to manipulate the poor immigrant job seekers. Thus, contemporary Arab cities, as will be explained, are parasitic, depending for their survival on the products of surrounding localities. This illogical and unacceptable human setting, although criticized, is largely accepted by many as other societal alternatives beyond capitalism (such as Islam) are not immediately available.

One might argue that today's Moslem cities are not Islamic, but rather cities inhabited by Moslems. For a city to claim the title 'Islamic', it should follow Islamic principles in its creation and formation. Those principles will create towns that are neither capitalist nor socialist, yet truly global in their character. To clarify this, we will explain in brief three Islamic principles: the first is the right of access to resources; the second is the concept of *'Ibn as-Sabil*, or the traveller; and the third is alms giving. These are only examples to bring the attention of both economists and sociologists to search for systems of rights (human rights, property rights, common laws, etc.) that transcend the failures of capitalism. One need hardly point out that the system of rights in any society is the major factor that shapes society's structure whether socially, economically, or physically.

In modern societies, access to resources, minerals for example, is limited to those individuals who have the means to obtain the required permission from the State. The system of rights in modern societies is much based on the vertical obedient relationships between individuals and the State. The State, for example, defines the common interest of the public for the public. If the State decides that such and such a mineral resource is quite rare and thus should not be exploited without its permission, it will have the right to limit access to that resource. This would often lead to the manipulation of that resource to the advantage of some over others. In Islam, however, the situation is drastically different. No one, including the State, has the right to limit any resource to any individual. Those who manage to reach any resource will have the right to exploit it. Of course, one would ask: but this would deplete some rare resources (oil for example)? It will, for sure, lead to the misuse of some minerals (uranium for example). Again, this is not the place to answer such concerns as it is beyond our context (the Arab Moslem city).[1] However, to understand the impact on the quality of cities of handing ownership of mineral resources (not lands) to those who can have access to them, we will first explain briefly the different opinions of Moslem scholars regarding access to resources. However, some explanation of the Islamic legal system is needed first.

The institution of Islam is based on three sources: the *Quran*, the traditions or sayings of the Prophet (*hadith*), and the teachings of jurists (*fiqh*). The first two sources were always referred to by jurists in reaching their rulings. This resulted in the development of different schools of law and gave the Islamic legal system its identity and cohesion. The most authoritative schools of law are: the Hanafi rite founded by Abu Hanifah (d. 150 H/767 AD) which covers India, Pakistan, Turkey, parts of Syria, South-East Asia and China; the Maliki school of law of Malik (d. 795 AD) which covers North and Central Africa, Upper Egypt, the Sudan and West Africa; the Shafi'i school of law of Imam Shafi'i (d. 820 AD) which covers Egypt, the southern and eastern Arabian peninsula, East and Meridional Africa and parts of South-East Asia; the Hanbali school of 'Ahmad b. Hanbal (d. 855 AD) which covers the entire Moslem world but does not prevail in any region except the central Arabian peninsula. Any individual can choose any rite or can even change from one school to another as they are all considered equally valid. The major differences between these schools are methodological, based on the particular method each founder used to interpret the two sources, such as *qiyas* (analogical reasoning), *ra'y* (opinion) and *ijma'* (consensus doctorum).[2]

The teachings of Islam regarding worship, or what relates a person to his God (prayer, fasting, etc.), were observed by most Moslems. However, what relates people to others, which is basically issues relating to rights, although explained by jurists at the theoretical level, was not fully implemented in practice except during the reign of the first four rulers (*Khalifs*). This is especially true regarding rights relating to the State's interest, for example in matters such as taxation. In general, early rulers were much keener to implement Islam compared to later ones. Thus, what will be explained next was never observed in our contemporary built environments, therefore current settlements cannot be characterized as Islamic, but rather as cities inhabited by Moslems.

Jurists have distinguished two types of mineral: surface minerals (*ma'adin zahirah*) and hidden minerals (*ma'adin batinah*). Surface minerals are the ones that can be obtained with little effort such as salt, while hidden minerals are those that need a great deal of effort to obtain such as gold. Hidden minerals were further divided into two types: solid minerals such as silver and liquid minerals such as oil. Further, jurists have classified properties that hold minerals into three types: private properties, treasury properties (owned by the State), and *mubahah* or accessible properties for the public.[3]

To minimize land speculation in order to maximize its utilization by as many individuals as possible to exploit minerals, the Islamic legal system has devised mechanisms to ensure that unutilized lands containing minerals cannot be owned by any party or individual, even by the State. To clarify this, we have first to explain the ownership of heights as an example.

Need and control without harming others have been the main prerequisites for establishing ownership. That is, only land which is needed and controllable can be owned. A debate took place regarding the ownership of what is beneath a territory. Al-Qarafi's opinion (from the Maliki school of law, d. 684 H/1285 AD) is that the owner of a territory usually benefits from heights for viewing rivers and gardens or for protecting his privacy by building parapets on his edifices, but such benefits do not exist beneath the ground beyond the foundation. Thus that which is beneath a territory cannot be owned.[4] This opinion was contested by 'Ibn ash-Shat, from the same school (d. 723 H), who pointed out that the owners of territories can, indeed, benefit from the ground by, for example, digging deep wells or basements. He argued that according to the principle of need there is no justification for preventing a person from deepening his well. Thus, the owner of a territory has the right to raise or deepen his territory as he wishes so long as he does not harm others.[5]

Regarding controllability as a determining principle of ownership, jurists debated the selling of the space on one's roof as a piece of land. Some schools of law consider the selling of height rights as selling the air above a territory, which is not controllable and therefore illegal. Other schools of law consider it as an ownership of the ceiling of the lower level which is controllable and ruled that an owner can sell the space on top of his house. Meanwhile, all schools of law agreed that an owner can sell the upper floor(s) or any part of his building – such as cantilevered parts – as long as it is built, since anything built is well defined and controllable.[6]

Thus, according to these two principles, no one can demarcate a piece of land to claim ownership if he does not need and thus use that land within three years.[7] However, according to the principle of revivification, any individual has the right to own unused land if he uses it by building on or planting it, i.e. occupying it by exerting effort. In other words, the only method to own land is by investing effort or capital. Thus, land appropriation by end users was common, since during the early Islamic period towns were expanding and land was often vacant. Not unexpectedly, appropriation has been extensively discussed by Moslem jurists. They recognized unowned and unused land as *mawat*, and followed certain principles in utilizing it.

Mawat literally means 'dead'. Land is considered *mawat* if there is no trace of building or cultivation; if it is not used by the neighbouring locality as, for example, a common pasture, burial ground, or as a source of minerals. According to custom, dead lands may be revived and consequently owned by the reviver. *'Ihya* (or revivification) literally means 'life-giving'; it means that controlling and using dead lands will bring ownership to the reviver.[8] There is ample evidence from the Prophet's traditions, rulers' actions and jurists' opinions to support the principle of assuming ownership of dead land by reviving it through cultivation or building on it.[9] Indeed, the concept of asking for permission from authorities is quite new in the Moslem world.[10] Owning dead land was a given right from the all mighty God.

If these principles are fully implemented, land with mineral resources will remain open for those who exert effort to obtain them. Cities will be characterized by dwellers who are often owners of the property they occupy, i.e. the city will be characterized by the least possible percentage of absentee owners. This will have tremendous impact on the quality of the city, socio-economically and environmentally. However, most importantly, as the State does not need and cannot utilize all available rural lands in the country, the State's ownership of land is reduced to the minimum compared to any other societal system (capitalism or socialism). Thus, land available for the access of end users to obtain minerals or to revive is the highest possible.[11]

Regarding minerals on land not owned by the State, the rulings of jurists varied if minerals appeared on privately owned land. Most Maliki scholars had the opinion that minerals on all lands, including private lands, are in the ownership of the State. The ruler should decide on their exploitation. This opinion, of course, resembles our contemporary system of rights in the Arab world. On the other hand, the jurists of the three remaining schools of law, as well as some Maliki jurists, although agreeing that minerals on private lands are not owned by the State, had two different rulings. The first is that of all Hanafi jurists, some Shafi'i jurists and Sahnun (a prominent jurist from the Maliki School of law, d. 854 AD) which states that ownership of minerals belongs to the owner of the land. The second ruling is that of most Hanbali jurists and some Shafi'i jurists which states that solid minerals should be owned by the land owner, however, liquid minerals should not be owned by the land owner. Yet, if such liquid minerals could be reached from neighbouring property, it should be owned by those who managed to obtain them.[12]

The jurists of the Hanafi, Shafi'i, and Hanbali rites agreed that land containing surface minerals could not be revived or owned and could not be allotted by rulers. The reason for such prohibition, they explained, is to facilitate public access to such minerals. Those minerals are just like water and pasture. They are for those who collect them. They argued that if such lands were owned, a class of wealthy people would be created who were not in fact productive.[13]

What if a group of individuals as partners managed to discover a remote site with hidden minerals, and invested much time by erecting the needed construction for mining, do they own that land, or do they have to give others the right to collect minerals from the same spot? Regarding hidden minerals in dead lands, the majority of rulings of jurists gave those who invested the right to remain in the area for mining, but that they should not hinder others trying to reach the same source of mineral from a neighbouring site. Jurists' rulings in these, and other, matters allowed for a maximum exploitation of resources by opening doors to as many end users as possible.[14]

As minerals are distributed naturally in remote sites, the Islamic legal system devised another principle that encourages travelling to mining sites. *'Ibn as-sabil*, or literally, the sun of the road, is a specific term for the individual traveller whose travel is funded by *zakat*, or alms giving.

Taxation in Islam, known as *zakat*, is generally 2.5 per cent of a merchant's annual income. If the person is a farmer, it is 5 per cent of the annual crop if the land is irrigated by human effort or 10 per cent if it is irrigated by rain. If the income is from mining, then it is 20 per cent of each extraction, etc. In fact, many details of *zakat* for all types of human activities were explained by jurists. What is interesting is that such taxation should be spent on eight types of people: the poor, the needy, *'ibn as-sabil* (the wayfarer or the traveller), etc. according to the verse:

> Alms are for the poor and the needy, and those employed to administer the (funds); for those whose hearts have been (recently) reconciled (to the Truth); for those in bondage and in debt; in the cause of Allah; and for the wayfarer; (thus is it) ordained by Allah, and Allah is full of knowledge and wisdom.[15]

From this verse, jurists insist that it is not even the right of the State to relocate the funds of *zakat* to any other project.

With this wealth of funds designated for *'ibn as-sabil*, jurists have developed an extensive literature on the conditions that give an individual the right to obtain funds to travel. If one reviews these conditions they are minimal, such as that a person will not be funded if he declared that he is travelling for a vacation, or if he is touring several regions. However, it is essential to favour *'ibn as-sabil* if he is seeking a job. In fact, *'ibn as-sabil* was mentioned seven times in *Quranic* verses encouraging Moslems to pay generously those wishing to travel. In one verse for example, the Holy *Quran* declared:

> And render to the kindred Their due rights, as (also) To those in want, And to the wayfarer; But squander not (your wealth) In the manner of a spendthrift.[16]

In other words, society has designated much of its resources to encourage job seekers, mineral explorers, and merchants to explore new sites for jobs, minerals and imports.[17] At the same time, Islam did not recognize borders between countries. The concept of a country demarcated by borders controlled by a State is quite modern in the Middle East. A world without borders would create a truly global situation that allows all members of society, especially the needy, to travel and exploit land resources. This is the exact opposite of our contemporary situation where borders between countries are opened only to the rich and the powerful to invest.[18]

Moreover, such travel would obviously be to sites of merit that contain minerals and other life supporting resources; i.e. to sites of productive jobs, and not to capital cities or regional cities occupied by bureaucrats as in our contemporary world. It is a well established fact that the current migration from rural areas to cities is due to several reasons among which is the presence of decision-makers in major cities. Most investments of contemporary societies (infrastructure, universities, hospitals, etc.) are in cities where the rich and powerful live at the expense of rural areas. Thus, hoping for a better life, migration of the poor has crowded cities as States have controlled access to resources. Although sites of contemporary cities do not support life, much of society's wealth is invested in them to resolve problems of over-population. Water is brought to Amman and Riyadh. Food is brought to Dubai and Casablanca. The location of modern cities does not reflect the merit of their sites, but rather the socio-political structure. Even mega cities, with merited sites, such as Cairo, are over-crowded and thus became consuming cities living on surrounding economies.

On the other hand, if the Islamic legal system is implemented, job seekers will be drawn to sites that need jobs and sustain life. A major principle in *zakat* is that it should first go to the local needy. The prophet, peace be upon him, has declared that the poor of each locality are more entitled to *zakat*.[19] This principle, coupled with the *'Ibn as-sabil*, would draw the poor to sites of the rich who can pay *zakat* and they, of course, are living in sites that have minerals and life support resources. Thus, sites of Islamic settlements compared to contemporary cities would reflect the merit of the sites and not the bureaucracy of the powerful. In such a scenario, settlements would compete to attract working people or job seekers. Over time, the unemployed (the poor) in crowded settlements would be attracted to newly established sites that are saturated by *zakat*, and so on. This mechanism would increase demand on workers. In such cases, the difference between the rich and the poor would certainly be much less than the disparity which the current globalization is producing. Thus we should expect a higher quality of life in such economically homogenous settlements, unlike the case of our contemporary globalized cities that are gradually transforming into a few islands of the rich in an ocean of the poor. Moreover, humans will not be forced to transport goods and life necessities from one site to another (which is costing societies much) as settlements are merited. Of course, not all sites could support all life necessities; some transportation of goods would be needed.

This flexibility of moving goods and people among sites to get the most out of our countries is reduced in this era of globalization compared to Islam, i.e., cities that are supposed to die are still living, while sites that are supposed to live are still covert.

In short, this brief prologue has tried to bring attention to the need for new concepts of rights that can support and reproduce today's advancements, beyond capitalism.

NOTES

1. Such issues are discussed in a forthcoming book by the author.

2. A fifth major school of law, the Shi'ite, which prevails in Persia, parts of Iraq and Lebanon and the eastern region of the Arabian Peninsula is not dealt with here. Neither did I investigate the Zaydi and Abazi rites which cover parts of the Arabian Peninsula. For details in English see Michon, 1980, pp. 21–22.

3. Properties were also classified differently regarding their ownership. The most well known classification that affected land subdivisions in the Arab world is the Ottoman classification made in 1858 AD which divided lands into five categories: (1) *Mamlukah* property, or property held by individuals in absolute ownership in which the owner could convert his property into a *waqf* (religious endowment) or bequeath it. Such actions were the highest form of manipulation, denoting a state of pure ownership. (2) *Miri* properties, or properties owned by the State and possessed by individuals who lived in them. (3) *Waqf*. (4) *Matrukah* property, or property left for public use. (5) *Mawat* or dead land. For further details see Akbar, 1988, pp. 47–52.

4. His argument is based on the principle that 'what is needed can be owned and what is not cannot' (*Wa ash-shar'u lahu qa'idah wahuwa 'annahu 'inna-ma yumlaku li'ajli al-hajah wa ma la hajata fihi la yushra'u fihi al-mulk*), Al Qarafi, Shihabud-din Abi Al-Abbasi As-Sanhaji, nd, Vol. 4, p. 17.

5. Moreover, he argued that if someone attempted to erect a room beneath his neighbour's territory he would unquestionably be stopped, even if he were to reach such a room from his own territory. Ibn ash-Shat, nd, Vol. 4, pp. 17, 40–41.

6. For details see Akbar, 1988, pp. 26–27.

7. Three years was the limit set by jurists to allow an individual to revive a land. For further details see Akbar, 1988, pp. 30–31.

8. Akbar, 1988, pp. 27–32.

9. The Prophet said, 'The people are God's people, the land is God's land, he who revives a piece of dead land will own it, and the unjust root has no right'. In another tradition he declared, 'He who revives dead land will be rewarded by God (in the day of judgment)'. 'A man who had revived dead land came to Ali (the fourth Caliph) and said, "I came across a land that was ruined or its (original) inhabitants had left it, and I dug streams and cultivated it." Ali responded, "Eat pleasurably (enjoy it) you are righteous not impious, a reviver not destroyer".' Ibn Qudamah relates that 'Reviving dead-lands is the custom in all regions even if there are differences among jurists regarding its regulation' (Akbar, 1988, pp. 28–33).

10. Most Jurists, with the exception of Abu Hanifa and some of his followers, agreed that revivification does not need permission from the authorities. For more details see Akbar, 1998, pp. 62–63.

11. For further details on these conclusions see, Akbar, 1998, Chapters 1 and 5.

12. Al-Abbadi, 1974, Vol. 1, pp. 349–351; An-Nawawi, 1995, Vol. 15, p. 232; Al-Mawardi, 1960, p. 198; Al-Hanbali, 1966. pp. 236; Ibn Qudamah, 1992, 15 vols. *Al-Mughni*, 1412, V. 4, p. 245; V. 5, p. 573.

13. *Al-Majmu*, v. 15 pp. 222–223, 227; ash-Shafi'i, Vol. V. 4 p. 42; *Al-Mughni*, V. 5, pp. 571–572; Al-Abbadi, V. 1, p. 355; V. 6, p 184.

14. The term 'end user' does not exist in traditional Arabic language as this class of people was created by capitalism. One might even argue that all people would be within the same class with minor economic variations if the Islamic Legal System was fully implemented. This utopian stratification and other related issues are discussed in a forthcoming book by the author.

15. *Holy Quran*: Translated by Abdullah Yusuf Ali; Surah 9 (at-Tawbah), verse 60.

16. *Holy Quran, op. cit.*; Surah 17 (al-Isra), verse 26.

17. *Holy Quran*, verses 60 of at-Tawbah, 41 of al-Anfal, 7 of al-Hashr, 177 of al-Baqarah, 36 of an-Nisa, 26 of al-Isra, 38 of Ar-Rum. *Tafsir at-Tabari* and *Tafsir al-Qurtubi*, verse no. 60 of surat at-Tawba; ash-Shawkani, V. 4, p. 169; *Al-Majmu*, V. 6, pp. 203–205 (printed 1415 H), *Al-Mugni*, V. 9, p. 330 (printed 1413 H).

18. Of course, questions regarding security will be raised. Again, it is not the place here to answer such concerns. Many similar issues are discussed in a forthcoming book by the author.

19. *Sahih al-Bukhari*, Tradition no. 1458 and 1395; ash-Shawkani, V. 4 p. 151–152; Abu Ubayd, *Al-Amwal*, Tradition no. 1908, 1911, 1916; *Al-Mughni*, V. 4, p. 131.

REFERENCES

Akbar, J. (1988) *Crisis in the Built Environment: the Case of the Muslim City*. Singapore: A Mimar Book.

Akbar, J. (1998) *Imarat al Ard fi al-Islam*, 3rd ed. (in Arabic). Beirut: Muassasat ar-Risalah Press.

Al-Abbadi, A.D. (1974) *Al-Mulkiyyah Fi al-Shariah al-Islamiyyah*, 3 Volumes (in Arabic). Published PhD dissertation. Jordan: Maktabat al-Aqsa.

Al-Hanbali, Abu Yala M. (1966) *Al-Ahkam al-Sultaniyyah* (in Arabic). Cairo: al-Halabi Press.

Al-Kasani (nd) *Badai' as-Sanai*, 2nd ed. 7 Vol.; Ala ad-Din b. Masud (in Arabic). Beirut: Dar al-Kutub al-Ilmiyyah.

Al-Mawardi, Abi Al-Hasan A. (1960) *Al-Ahkam al-Sultaniyyah* (in Arabic). Cairo: al-Halabi Press.

Al-Qarafi, Shihabud-din Abi Al-Abbasi As-Sanhaji. (nd) *Al-Furuq*, 4 Volumes (in Arabic). Beirut: Dar al-Marifah Press.

An-Nawawi, M. (1995) *Kitab al Majmu* (edited by Al-Mutii, Muhammad N.) 23 Volumes (in Arabic). Jeddah:

Ash-Shafi'i, Muhammad b. Idris (nd.) *Al-Umn* (edited by M. Zahri an-Najjar) (in Arabic). Beirut: Dar al-Marifa Press

Ash-Shawkani (nd) *Nayl al-Awtar*, 4 Volumes (in Arabic). Beirut: Maktabat ad-Dawah al-Islamiyyah

Ibn ash-Shat (nd) *Hashiyat Ibn as-Shat ala al-Furuq*, 4 Volumes (in Arabic). Beirut: Dar al-Marifa.

Ibn Qudamah, Abi Muhammad A. (1992) *Al-Mughni* edited by al-Turki, A. and al-Hilo, A. (in Arabic). Cairo: Hajr Press.

Michon, Jean-Louis (1980) Religious institutions, in Serjeant, R.B. (ed.) The *Islamic City*. Paris: UNESCO, pp. 21, 22.

Chapter 3

The Spatial Development and Urban Transformation of Colonial and Postcolonial Algiers

Karim Hadjri and Mohamed Osmani

Introduction

Algiers is a complex and confused city that is experiencing constant change at the administrative, political, social and economic levels. Its administrative status changed in 2000 from that of a region managed by a Governor Minister to that of a simple *Wilaya* (County). This change has meant that the new plan proposed by the former Governor of Algiers – the *Grand Projet Urbain* (Major Urban Plan) – that would lead to the establishment of Algiers as an efficient metropolis, a regional hub or even a global city, is on hold. That said, for Algiers to achieve the status of a global city, it must meet several criteria, which up to now seem far from being part of its perceived growth and objectives. Nevertheless, before embarking on the analysis of Algiers's development history and current challenges, there is a need to define the central topic of this chapter, which is the urban transformation of Algiers during and after colonization.

This chapter sets out to explore the spatial development and urban transformation of colonial and postcolonial Algiers by examining the pre-colonial *medina* or Casbah. This exploration reveals the complexity of the city's development, its regional and international position and influence, and the extent to which globalization has affected its urban growth and architectural patterns.

Algiers presents at least six stages of urbanization that produced distinct urban fabrics, namely, the Casbah; the colonial town (Algiers centre and surrounding neighbourhoods); colonial *Grands Ensembles* (large urban housing projects); post-colonial *Grands Ensembles* or ZHUN (*Zone d'Habitat Urbaine Nouvelle*); urban development by private builders on individual plots of land; and informal housing (*bidonvilles* or shanty towns and housing of the poor). It should be noted that Algiers's early morphology, defined by the Casbah and the colonial quarters, has been retained.

This chapter will examine the spatial development of Algiers by concentrating on two phases of its urban growth and transformation, the colonial period, from 1830 to 1962, and the postcolonial period covering 1962 to the present day. The influence of colonization on urban form and architectural trends will be emphasized. In addition, the effect of globalization on post-colonial Algiers and its impact on local identity and urban and architectural models will be discussed. Algiers's most complex urban issue is the restoration and safeguard of the Casbah. A brief account of the Casbah's decay and renovation projects will also be examined throughout the chapter.

On 21st May 2003, Algiers and its surroundings experienced a devastating earthquake that put to the test both old and new buildings. The impact of this disaster on urban development actors and existing urban tissue so far will be briefly discussed.

Colonial Algiers

Early records of Algiers date back to the fifth century BC when the site attracted the Phoenicians. After this, no major events were recorded until the first century AD when Algiers became a Roman colony. Berbers inhabited the region of *Sahel* for centuries and records of these settlers date back to the first century AD (Chennaoui, 2000*a*). The Arab period had two stages: the first from the sixth century after the collapse of the Roman empire when there were the successive attempts by the Arabs to invade the settlement, which led to the destruction of the Roman buildings; and second, the founding of Algiers (*El Djazair*) by Prince Bouloughine Ibn Ziri around the middle of the tenth century. During the Middle Ages, the Europeans attempted to invade the city several times. The fifteenth century saw successful settlement by the Spanish who started by fortifying the city. In 1518 the Spanish were replaced by the Turks under the Ottoman naval command of the Barbarosa brothers who became the rulers of Algiers. During the sixteenth century, the city was further fortified and developed by building ramparts and citadels and a seaport (Chennaoui, 2000*b*). Algiers's economy blossomed under Turkish rule mainly within the Casbah, which itself is inspired by Turkish vernacular architecture, socio-cultural and religious requirements, and the particular characteristics of the site. Some aspects will be examined in detail throughout the chapter, but this section concentrates on the French occupation of Algiers, dating from 1830.

During the war-torn eighteenth century in Europe, the French found themselves in great need of agricultural products. Algeria, a rich producer of cereals at that time, supplied France with large amounts of wheat, with a long-term credit facility. Quarrels over settlement of these debts helped give rise to a turbulent relationship between France and Algeria (Laffont, 1981). Specifically, the French government did not answer letters from the Dey (Governor) of Algiers, concerning the credit. It was within this climate that the famous 'fan scene' took place. In 1827, the Dey Hussain received the French consul, Pierre Deval, and blamed him for the silence from Paris (Renaudot, 1979). Deval replied in offensive terms, which resulted in a violent act from the Dey who hit the consul with his fan. This incident was the driving force behind a political crisis between the two countries. The refusal of the Dey to offer an apology appeared *prima facie* the motive for the invasion of Algeria. The conquest of Algeria was also to form a publicity coup for France, making it a powerful country competing with other European states. The French army occupied Algiers on 5 July 1830; this marked the beginning of a colonization, which lasted 132 years. During this period, the Casbah of Algiers was subject to profound social, economic, cultural, and physical-spatial changes (Osmani, 1988).

The case of Algiers provides the best example for understanding the colonial urban practices, which were reproduced in a variety of forms, in other Algerian cities. Within the broader colonial urban history of Algiers, five distinct periods may be identified:

- 1830–1840: military planning
- 1840–1880: the birth of the European city
- 1880–1914: extramural expansion and urban growth
- 1914–1945: transformation between the two world wars and rural migration
- 1945–1962: housing shortage and *bidonvilles*

1830–1840: Military Planning

A series of radical transformations, which occurred between 1830 and 1840, had such a wide-ranging impact that any attempt at present day reconstruction of the Casbah of Algiers has been extremely difficult. The initial colonial urban intervention in Algiers was characterized by military engineering, which 'dominated all urban design operations during this period' (Çelik, 1997, p. 27). This military supremacy was stated in a report from this period, 'the organization of space in Algiers must submit to the changing of institutions; the Christian commercial town cannot preserve the form of a pirate capital. A transformation is imminent' (Miege, 1985, p. 173). Some military officers even declared their desire to pull down the whole lower city and rebuild it in the French manner (Lespes, 1930). However, the costs of expropriation and the protest of the indigenous people made it difficult to achieve this aim.

The first decade of colonization was characterized by a cultural approach and a military need. Culturally, the French colonial policy 'sought to sanitize what was perceived as unsanitary' (Stewart, 2001, p. 178). Militarily, there was a compelling need to reuse pre-colonial structures for accommodation and logistical purposes (figure 3.1). The troops occupied the Casbah, forts, batteries, and many houses near the ramparts as well as several mosques. Outside the walls, the *funduks* (inns and warehouses) and palaces were converted into military bases.

As a military city, Algiers needed wide roads and large squares for the rapid circulation of troops. Fulfilment of these requirements came initially in road construction within the walled city. Between 1830 and 1840 the three main pre-colonial streets of Rue Bab Azzoun, Rue Bab el Oued and Rue Bab el Ghazira (renamed Rue de la Marine) were widened, at the expense of the existing houses, shops and public buildings. While the original width of streets in the Casbah had been conceived to allow two fully loaded donkeys to pass, it was now determined according to the gauge of military vehicles (Osmani, 1988).

The transformed streets were lined with two- and three-storey, arcaded buildings built for the new European immigrants. Retailing and commercial functions were concentrated at ground level and the two floors above were given over to flats. Uniform façades characterized the construction along these streets. The arrangement of windows and balconies further emphasized the vertical regularity of the façades. Thus, with this initial transformation, the traditional street of Souk el Kebir (the Great Market) disappeared and was replaced by a monumental street linking the east and west military forts and allowing soldiers a clear space for parades.

During the same period, a large square, Place d'Armes (initially called Place Royale then Place du Gouvernement), was created at the meeting point of the three streets (Rue Bab Azzoun, Rue Bab el Oued and Rue de la Marine). The Place du Gouvernement, one side of which opened to the sea, took a rectangular shape and was framed by three-storey, arcaded buildings (figure 3.2). The

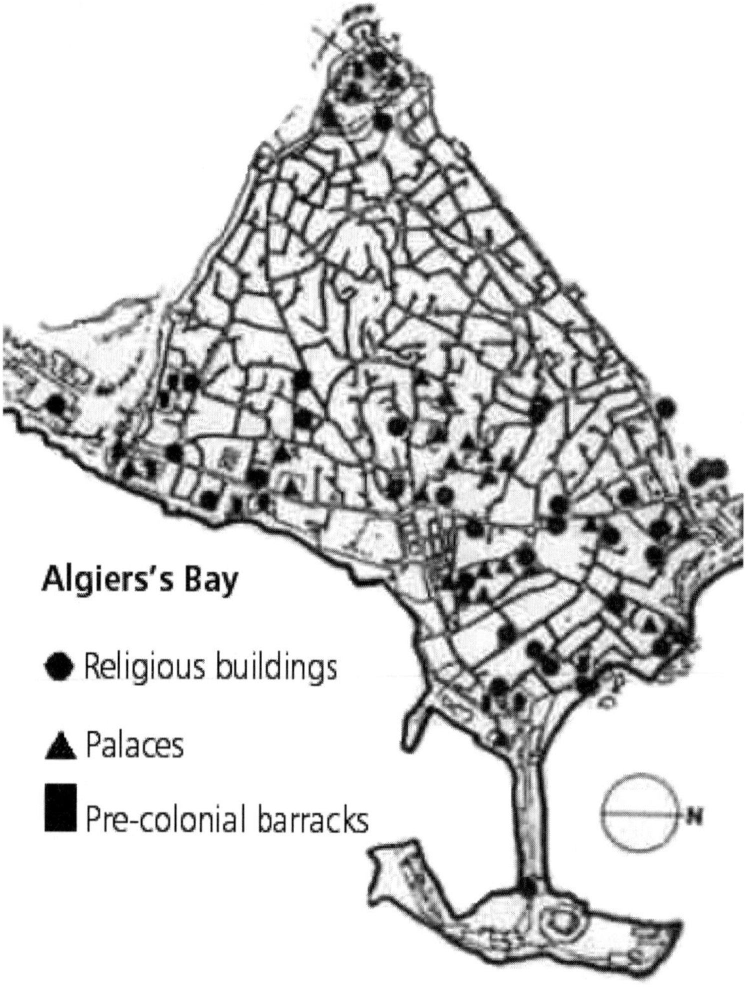

Figure 3.1. Pre-colonial buildings used by the French army between 1830 and 1840.

Hotel du Gouvernement, which faced the sea, was erected as the focal point of the whole spatial organization of the new square. The Place du Gouvernement reflected urban design principles developed in France during the eighteenth century for the planning of the Places Royales: a square connecting long axes and so giving a sense of space and openness (Osmani, 1988).

After the modifications of the original urban structure, the indigenous population, which felt that its privacy was violated, moved away from the lower city to the upper city or to rural areas; their departure made room for the new European arrivals. By 1839, 15,000 Europeans (French, Italians, Spanish and Maltese) were concentrated in the lower part of the city. Between 1830 and 1840, 20,000 Algerians left the city and were replaced by the same number of Europeans (Lespes, 1930). Some of the newcomers occupied traditional houses that were subsequently transformed with large windows opening onto streets and by covering the courtyards (Institut Français d'Architecture, 1984). Hence, the original built form of traditional houses was trans-

Figure 3.2. Place du Gouvernement, Algiers.

formed by accommodating the habits of the new occupants.

Many traditional houses were destroyed and replaced by blocks of housing. Plot after plot and block after block, a system of secondary streets was created, thus forming a new urban structure. The façades, widely opened onto the streets, and the extension of dead-end streets to serve and link the blocks, was in sharp contrast to the pre-colonial organization of the city, characterized by closed residential quarters. Therefore, a new system based on continuity and the opening of streets interrupted by open spaces was superimposed.

1840–1880: Birth of the European City

The continual arrival of European migrants, estimated at 21,000 in 1841, accelerated the rate of urbanization in Algiers (Comedor, 1972). The need to expand the town outside the walls became an absolute necessity. In 1840 the process of development of the city began with the creation of a new urban fabric that was no longer superimposed or reclaimed from the pre-existing urban fabric. Nonetheless, the latter was not left untouched; for security reasons and because most of the colonial institutions were located in the lower city, the Casbah of Algiers had to be connected to the new urban development by a web of streets and squares. This new linkage system changed the original form of some traditional parts of Algiers's Casbah. Outside the walls the army seized almost all the flat sites and monopolized extramural construction. Consequently, military construction created a physical impediment constraining any process of urban development for the civilian population.

In 1841, the colonial administration established a development plan to allow the expansion of the city outside the original nucleus (Kecir, 1980). To accommodate the new European city, the construction of new fortifications started in 1841 and was completed by 1848 (Royer, 1932). The army opposed any civil construction and was supported by the French central government in

1844 which stated that, 'Algiers must be a war place, which forms the citadel of Algeria, rather than a commercial city' (Royer, 1932, p. 33). After lengthy discussions between the military and civil administrations, the general development plan for the expansion of the city was finally approved in 1846. The layout of new streets and open spaces was the main concern of this plan. By 1849, the old Turkish fortifications were 'replaced by new ones, enclosing an area three times larger than the one occupied by the old town' (Çelik, 1997, p. 35). Thus, the main concern was for salubrity, aeration and security. Gardens with a system of slopes and steps were organized in places where declivity was important (Institut Français d'Architecture, 1984).

All projects of construction or embellishment had to proceed through endless administrative routes. For example, even the pavement of several streets, such as the Rue de la Marine, had to be approved by the Minister of War in Paris (Lespes, 1930). Such bureaucratic processes together with the previously mentioned military monopolization of the flat sites, retarded the urban development of Algiers.

In 1850, the creation of the Place Bresson (today Place Port Said) on the site of two small old squares, Place de Garamantes and Place de Bourneau, bridged the physical gap between the old city and the new European quarter (figure 3.3). In fact, details of the new square and concepts used to link the old urban fabric with the new urban development were almost exact copies of the design of the Place Royale in Nancy (Osmani, 1988).

However, the general development plan was not totally achieved. Part of the programme was executed, the rest was started belatedly then forsaken for a number of reasons, particularly the 1846 economic crisis.

The number of houses constructed in the new European quarter was not in proportion to the number needed. While construction escalated, the population remained almost constant (Lespes, 1930). Individual buildings were being sold with difficulty. Insolvencies increased and speculators preferred to invest in the railway construction in the metropolis. Furthermore, the financial crisis in France at that time affected the spatial development of Algiers, as more than a quarter of the European population left the city. Indeed, the foreign population decreased from 42,000 in 1846

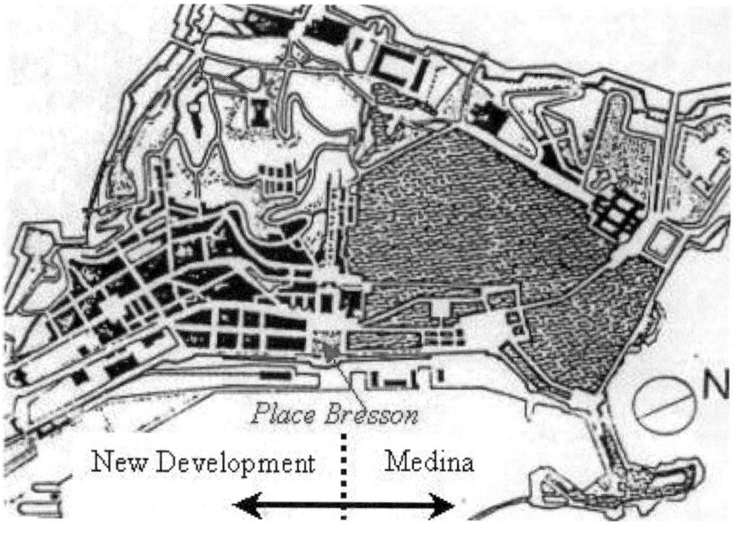

Figure 3.3. Plan of Algiers in 1880.

to 25,000 in 1853 (Comedor, 1972). In addition to the economic and financial crisis of 1846, a political crisis broke out in France in 1848 with the fall of the monarchy.

To the south of the Casbah, the less sloping sites along the coast were reserved for the army. Higher up, a new residential and administrative quarter, the Quartier d'Isly, developed. Two unequal urban structures separated by a large artery, the Rue d'Isly (figure 3.6, p. 38), characterized the physiognomy of the new quarter:

- a low and large part in which streets crossed each other at right angles round a square, the Place d'Isly;
- beyond the Rue d'Isly, streets were sinuous bending into a broken slope.

The Casbah of Algiers was clearly separated into two distinct parts by the Rue Bab Azzoun, the Place du Gouvernement and the Rue Bab el Oued. The first part consisted of a flat area below the principal artery where the main European institutions were located. These institutions included the Governor's Palace, courthouse, theatre and various public buildings. This part was gradually depopulated of its Algerian residents and became exclusively European. The quarter formed by the Place du Gouvernement, the Rue Bab el Oued and the coast, had a more regular pattern than the rest of the old city.

The second part was the slopped district, El Djabel (the mountain), where the indigenous residential quarters terraced up to the citadel. This part of the city kept its traditional urban form. A twisting ramp, Montée Ravigo, connected the upper town with the new European quarter.

The block formed the basic structural element of the new city; the urban structure was conceived as a combination of rectangular, square and triangular blocks. The residence-blocks and monument-blocks characterized the physical form of the new urban quarter. In fact, the same kinds of blocks were laid out for the reconstruction of Paris, where Haussmann reserved specific blocks for monumental buildings such as Notre Dame and Hôtel-Dieu (Osmani, 1988).

The expansion of Algiers southwards brought about the development of the harbour installation. The port was to become one of the most important economic elements of Algiers. It was first extended in 1857 in response to the establishment of railways in the colony. Algiers's port became a fundamental link between the Algerian hinterland and France.

When Napoléon III came to power in 1852, a new impetus was given to the programme of expansion and embellishment of Algiers. According to Lespes (1930), Napoléon's personal views concerning the situation of the city were threefold:

- The priority of French colonization was to be given to commercial and industrial activities.
- The extent of military institutions, which were impeding the urban development of the city, was to be limited.
- The continual destruction and transformation of the indigenous quarters in the upper town were to be ended.

The Boulevard Front de Mer, later called Boulevard de la République, was the most significant achievement of the Napoleonic regime. Fronting the sweep of the Bay of Algiers, the Boulevard de la République was lined with buildings characterized by regular façades with arcades at the ground floor level and, above, three or four storeys with French windows opening onto continuous balconies (figure 3.4).

The horizontal lines of the façades on the Boulevard de la République emphasized the monumentality of the architectural composition, reflecting homeland trends. 'The French architects and urbanists of the later 19th century turned, automatically, to the architectural style with

Figure 3.4. Boulevard de la République (today Boulevard Che Guevara), Algiers.

which they were familiar back home in Paris and major provincial cities' (Davis, 1983, p. 15). Similar architectural features could be seen on the Rue de Rivoli in Paris, to which the Boulevard de la République bore close resemblance.

As a result of the economic crisis the programme for Algiers's expansion saw a stagnation that lasted several years. It was not until 1855 that different projects for the urban development of the city were proposed. Three projects designed by Vigoureux and Callat, MacCarthy and Genevay, and Chausseriau were taken into consideration. All of these proposed the creation of a new city outside the walled area. If the reasons justifying this new idea were numerous, their common denominator seemed to be the impossibility of adapting the indigenous city to the needs of the French colony. Most significant was the project of Chausseriau designed in 1858.

Chausseriau's ideal of the new city was based upon hygiene, comfort and aesthetics (Beguin *et al.*, 1983). To achieve his aims, he proposed a city plan for 60,000 inhabitants with large axes, numerous gardens, fountains and squares. Medium-rise buildings provided with spacious galleries would frame the streets and open spaces.

1880–1914: Extramural Expansion and Urban Growth

Starting in 1880, wine production was the main driving force of colonial urban development in Algeria. After the total destruction of the vineyards in France by *Phylloxera* in 1875, the French government encouraged large-scale planting of vineyards in Algeria (Comedor, 1972). For better exploitation of the hinterland, a railway network was developed linking the rural areas with different ports, so enabling the export of agricultural products to metropolitan markets. Therefore, colonization, development of cultivation, the increase in wealth, and urbanization were all closely linked.

The success of viticulture resulted in an economic 'boom'. Algiers, as the administrative capital, took the lion's share of this economic effervescence. The port was enlarged and new administrative and commercial institutions were established. The population increased considerably. It jumped from 84,000 inhabitants in 1880 to 172,000 in 1911 (Miege, 1985). In 1881, the population of Algiers was distributed as follows: 46,482 inhabitants in the old city; 6,061 in Bab el

Oued; 13,050 in the Quartier d'Isly; and 12,279 in Mustapha (Comedor, 1972). This distribution of population shows that the development of the city originally took place towards the area situated south of the old core, rather than the northern part where the Bab el Oued quarter was of limited importance. Housing construction became one of the most reliable and profitable investments. The establishment of the tramway in the 1880s helped to link the northern and southern suburbs with the city (Institut Français d'Architecture, 1984).

In 1891, after lengthy discussions between the military authorities and the municipality, a series of concessions concerning the southern and northern sites was obtained for the expansion of the city. The main hindrance, fortifications and military zones, which stood in the way of the physical integration of the intramural city and the new quarters, was finally removed. Between 1880 and 1914 the city burgeoned, stretching southwards along the sea (figure 3.5).

The plan of embellishment and expansion of Algiers imposed certain modifications on the urban structure of the Casbah by creating new streets in the upper city, linking the indigenous city and the European quarters.

The most remarkable development was towards the south. Here, the main focus was the creation of the commercial centre of Algiers. This had to be realized out through the development of the port and the construction of the railway station, harbour and warehouses. Large arteries connected this new centre with the rest of the city. The northern part of the city developed from scattered constructions to an industrial zone. Avenues 12, 18 and 20 metres wide were laid out and flanked by arcaded buildings (Institut Francais d'Architecture, 1984). A large artery, the Boulevard du Général Farre (now Boulevard Abelrahmane Taleb) was constructed on the ruins of the old fortifications. It extended from the sea up to the citadel, and was intersected by several streets connecting the northern part with the centre.

Attesting to the gradual displacement of the hub of the city towards the south, two monumental buildings in neo-Moorish style were erected: the Prefecture (today the Wilaya) along the sea front and the main post office, articulating the city with its suburbs and emphasizing the location of the new centre.

The image of the city changed. Gradually, old buildings with plain façades were pulled down

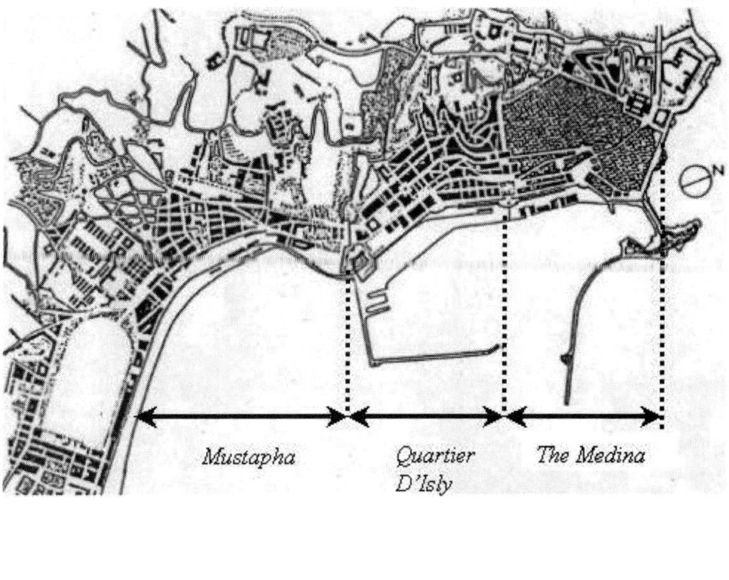

Figure 3.5. Plan of Algiers in 1895.

and replaced by new ones incorporating balconies and bow-windows (figure 3.6).

The architectural aspect of Algiers was firmly established by the end of the nineteenth century. In some cases, the topography of the site contributed to the variety and multiplicity of building types. In short, the idea of Algiers as a military centre and a naval port had become completely obsolete and the capital of French Africa developed new functions and became a commercial emporium.

1914–1945: Transformations between the Two World Wars, and Rural Migration

By the beginning of the twentieth century, Algiers's urban structure was almost in its final form, characterized by a sharp dichotomy between the Casbah and the new European quarters (figure 3.7). The peripheral quarters were articulated around the new city centre surrounding the main post office.

The effects of the First World War were felt in many aspects of Algiers's urban situation, including a decrease in European immigration, stagnation of construction, and the postponement of the enlargement of the port.

In 1930 the colonial administration instituted a law that, even today, rules most Algerian urban centres. This required the formulation of a *Plan d'Urbanisme* (general development plan) for any city of more than 10,000 inhabitants (Deluz, 1980). The first *Plans d'Urbanisme* that were prepared showed a transition from the alignment planning of the nineteenth century to the zoning of the twentieth century.

Two major phenomena characterized the period preceding the Second World War:

- The appearance of 'modern' architecture in Algiers. Indeed, many buildings such as the Palais du Gouvernement were erected reflecting new architectural trends.
- The emergence of the first social housing in Algeria. Poor Europeans (Italians, Spanish and Maltese) lived in unhealthy houses in semi-industrial suburbs and inside the Quartier de la Marine.

Figure 3.6. Rue d'Isly, Algiers.

Figure 3.7. Contrasting urban fabrics: Casbah and colonial urban development, Algiers.

Furthermore, during the inter-war period, the high residential density in the Casbah forced some inhabitants to move towards the suburbs, forming spontaneous and anarchic constructions. To solve housing problems, *l'Office Public des Habitations à Bon Marché – HBM* (Public Office for Low Cost Housing) was created in 1921 (Deluz, 1980).

The municipality decided on the destruction of the insalubrious Quartier de la Marine. In 1940, almost all the old buildings in the area were razed and a new programme of redevelopment was launched. The local authorities appealed to many architects and planners for advice. These included Henri Prost, who designed most of the master plans for the Moroccan cities, Jacques Gréber, Maurice Rotival and finally Le Corbusier (Emery, 1980). 'The problem was to define an order capable of growth and change, but with certain fixed points of monumentality appropriate to Algiers's increasing importance in the French colonial order' (Curtis, 1986, p. 122). Le Corbusier arrived in Algiers in 1931; he visited the Casbah as well as other Algerian *medinas*, such as Ghardaia (Miquel, 1980). In his proposals, known as the *Plan Obus*, Le Corbusier conceived four main elements 'orchestrated into a sculptural whole' (Curtis, 1986, p. 123):

- a monumental skyscraper for a new business centre intended to regulate the scale of the city;
- a structure of concave and convex apartment buildings, on the model of *l'Unité d'Habitation* to be built for the middle class;
- an elevated roadway running north-south above the Casbah;
- a serpentine viaduct for working classes, running along the bay, and providing walkways and shops.

The *Plan Obus*, 'the new bombardment of Algiers's' as some observers called it, was a reflection of *ville radieuse* principles, intended to fit into different cultural and environmental contexts. Le Corbusier's proposals for the redevelopment of the Quartier de la Marine were purely utopian, and his project was not realized. Instead the site saw a huge development of a speculative nature. High- and medium-rise blocks were erected facing the sea. This new concrete screen blocked the views to the bay, which the pre-colonial Moorish houses had enjoyed. This last operation speeded the decay of Algiers's Casbah, which subsequently became an overcrowded ghetto.

The new nature of the colonial economy encouraged European immigrants to settle in the countryside. Their settlement was facilitated by a law codified in 1863 which permitted property sales to Europeans (Lawless, 1979). The newcomers appropriated vast fertile lands, given over to cash crops relevant only for export and foreign consumption. 'Squeezed off their traditional lands and experiencing population pressure as a result of galloping demography, a growing number of Moslems were forced to seek jobs as laborers on European farms or in the mines' (Lawless, 1979, p. 83). Looking for a better income, rural and nomadic people from the interior were drawn towards the urban centres. 'They provided a cheap labor pool used to construct the housing and infrastructure of the new cities, to man factories capitalized from abroad and oriented toward Europe, and to serve the imported "elite" as domestics, gardeners, and porters' (Abu-Lughod, 1976, p. 193).

There was a striking increase in the Algerian population in the big cities between 1930 and 1960. Algiers saw its population grow from 100,000 to more than half a million people in 30 years (Lawless, 1979). The continual influx from the rural areas was particularly speeded by the establishment of industries in these cities. High-density Moslem populations were found around the major coastal cities. These cities were socially and economically unable to cope with the continual internal migration. From this situation two major problems arose. Firstly, the interior towns became resource centres with exclusive focus on the metropolis. Secondly, the high population of the major cities created many economic and social problems, such as the high rate of urban unemployment among the Algerian population and housing shortages. 'Poorly integrated into the colonial urban economy, the new Muslim proletariat lived apart from the European residents – separate as well as unequal' (Lawless, 1979, p. 84). This was particularly true in Algiers where the Casbah concentrated the early migrants. By 1949, it became rapidly overcrowded, reaching 3,500 inhabitants per hectare (Çelik, 1997).

1945–1962: Housing Shortage and Bidonvilles

The post-war period did not emerge with any evident changes. Unlike the European experience of bombing and destruction necessitating huge programmes of reconstruction and development, the urban situation of Algiers seemed unchanged (Osmani, 1988). The same architects continued to produce the same type of architecture and the city slowly expanded towards the south and south-west.

The year 1954 was a key date in the history of Algeria and specifically Algiers. An earthquake hit Orleanville (today Chlef) destroying the city, ironically acting as a catalyst for a new approach. In the same year two political-economic determinants in the history of modern Algeria emerged: the outbreak of the Algerian War of Liberation and the discovery of oil.

The 1950s saw a decisive reversal in Algiers's population growth. The Algerian population, a minority until the beginning of the War of Liberation, experienced a major increase. In 1906 it represented no more than 23 per cent of the total, by 1954 it made up 51.4 per cent of the whole population. Thus, in one generation, the Algerian 're-conquest' of the city occurred. The 'population explosion' was chiefly the result of massive migration from the countryside, strongly felt after the war. The first newcomers settled in the Casbah, 'filling the *funduks* and *wakalas* [dormitories for itinerant merchants and travellers in the early days] and then piling up in the existing structures' (Abu-Lughod, 1976, p. 208).

The continuous pressure on dwellings led to the rapid overpopulation of the Casbah, which could no longer absorb the accelerating rural flow. Most of the later migrants settled in spontaneous quarters, consisting of self-built housing created using any available materials. These squatter settlements, which developed primarily on the periphery of the big cities and in some cases reached the heart of modern quarters, were known as *bidonvilles* (shanty towns).

In Algiers, there were only a few *bidonvilles* around the European quarters, whilst they mushroomed in and around the Moslem quarters. As noted by Miege (1985, p. 178), 'a high proportion of Algerians lived in these *bidonvilles*, as opposed to half in the Casbah, ten percent in Bab el Oued and less than five percent spread over the remainder of the town'. The number of the Algerian population living in *bidonvilles* jumped from 85,500 inhabitants in 1954 to 144,000 in 1962 (Institut Français d'Architecture, 1984).

Thus, the number of *bidonvilles* increased dramatically, occupying unbuilt areas of the city and many sloping sites. These scattered areas cramped the urban development of the city. To combat this spatial problem, the municipality worked out a programme for housing development in which two separate operations were launched (Institut Français d'Architecture, 1984). The first concerned the construction of new quarters for Europeans resembling the big housing estates on the outskirts of French cities, though the green spaces and the different facilities originally planned were never realized. The second operation was associated with the creation of *cités de recasement* for Algerians, in which even minimal facilities required were not conceived – no asphalt for roads and footways, no facing for façades and an area of just 30 to 35 square metres for any dwelling.

With the intensification of the War of Independence, the programme of housing development was an ultimate attempt by the French to retain their position in Algeria. 'Attempts to re-house some of the occupants of the *bidonvilles* during the last decade of the colonial rule, as part of the counterinsurgency measures adopted by the French, represented a classic case of too little too late' (Lawless, 1979, pp. 84–85).

In 1959, a large-scale economic and social development plan was launched. It was the major cities which largely benefited from the investments of the Constantine Plan whose objectives were twofold:

- the long term with a policy of regional planning, with special interest in the creation of new industrial enterprises;
- the short term with development of housing estates for both French and Algerian populations.

The plan also developed a set of guidelines concerning the industrialization of the interior towns. These industrial zones, which were conceived to accompany important housing programmes, formed *cordons sanitaires* and *remparts urbains* so as to avoid the overpopulation of European urban centres by rural migrants (Mutin, 1982).

The Constantine Plan was a short-lived programme due to the process of de-colonization, which began with Independence in 1962. Nevertheless, its legacies were strongly felt during the first decades of Independence. Almost all the projects, which were not finished during the colonial period, were completed later and the main objectives of the plan were, in general, respected in the postcolonial development of Algerian cities. By the end of colonial rule, almost all the colonial cities showed the same spatial organization: industrial zones, areas of social facilities, administrative sectors, and commercial quarters.

The extended growth of Algiers and the large, economic and social projects did not solve the problem of housing shortage. On the contrary, the situation was worsened with the installation of industry in the cities attracting more rural migrants.

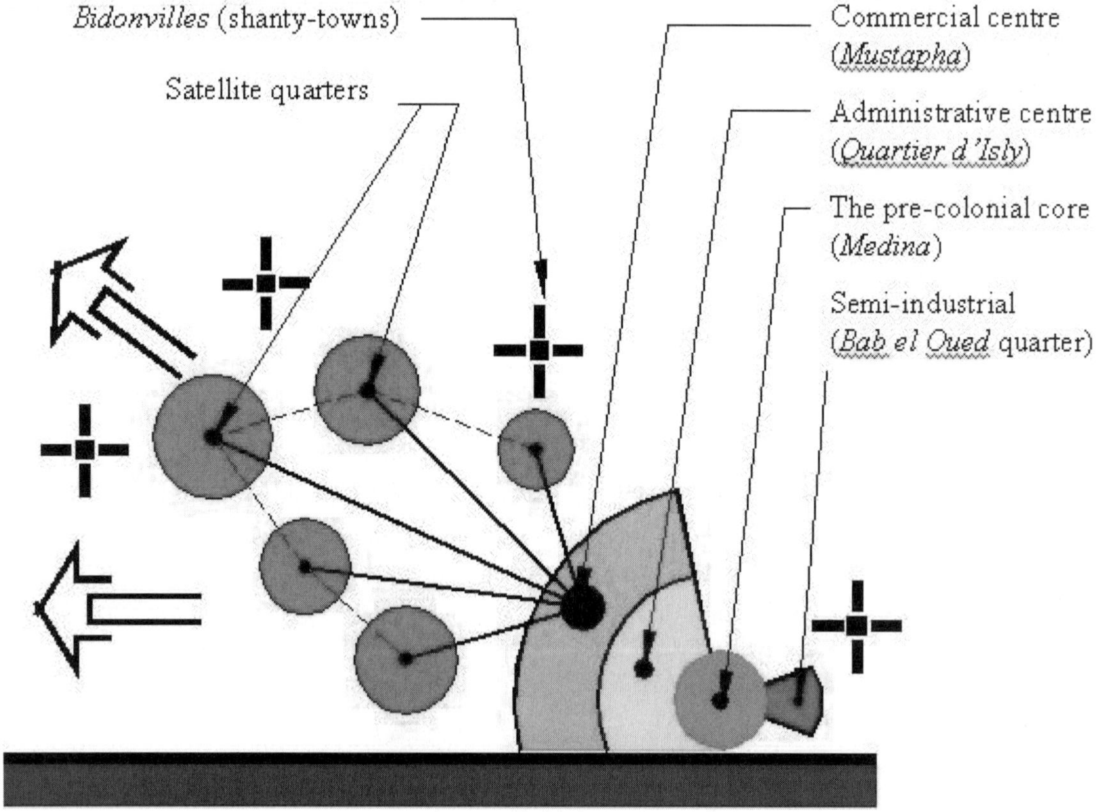

Figure 3.8. Spatial development of Algiers during the colonial period: 1830–1962.

On the eve of Independence the plan of Algiers showed three antithetical patterns:

- the dense urban fabric of the Casbah;
- the grid structure of European quarters;
- the scattered *bidonvilles.*

In summary, the Casbah of Algiers, unlike the pre-colonial cities in Tunisia and Morocco, was subject to considerable physical and functional alteration and distortion. After a long period of settlement, characterized by military intervention in which new streets were cut through the old urban fabric, squares created and multi-storey apartment blocks erected replacing traditional houses, new extensions occurred north and, especially, south of the old nucleus. The northern part of Bab el Oued was developed as a semi-industrial sector populated chiefly by poor Europeans (figure 3.8).

The southern expansion, Quartier d'Isly, constituted the first part of the 'dual city' in which the main administrative and cultural activities were concentrated. A further development, which stretched southwards and in which commercial functions were assembled, emphasized the dichotomy of the two urban structures.

Several satellite quarters such as Belcourt, Hussein Dey and Kouba were articulated around the administrative and commercial centre. Each quarter embodied a central core of activities and European residential areas. Around some of these urban units industrial activities developed, attracting migrants from the countryside. Without any spatial organization, the newcomers settled in and around peripheral quarters forming *bidonvilles*, the most enduring colonial legacy, impeding the spatial development of the city towards the south and south-east. A system of streets, linear whenever possible, was established to link the satellite quarters with the centre.

The next section will discuss urban growth of postcolonial Algiers, and urban planning and development challenges facing contemporary Algiers.

Postcolonial Algiers

Having discussed colonial Algiers, the second part of this chapter will concentrate on the post-colonial period. The city's urban growth and development patterns are examined in order to assess the effects of globalization on the city and its impact on local identity and urban and architectural models.

Given that architecture and urban form are important elements that define local culture and identity, emphasis will be on the influence of colonization and its legacies on the growth process of Algiers. There is no doubt that colonization in particular had tremendous effects on social and cultural perceptions of the colonized.

The aim here is to examine the postcolonial period through five stages that fortuitously fit within decades, because most crucial events such as the 1965 *coup d'état*, the death of President Houari Boumediene in 1978, dramatic decrease in oil revenues in 1986, and social unrest and terrorism since 1988, happened during the second half or at the end of each decade.

These are:

- Colonial continuity: 1960s
- The Socialist Period: 1970s
- The Transition Period: 1980s
- The Free Market Period: 1990s
- Algiers in the New Millennium

Colonial Continuity: the 1960s

This period starts with the Independence of Algeria in July 1962. City planning discourse at the time concentrated on the renovation of the Casbah, the housing shortage due to rural migration, and the *bidonvilles*.

During the early years of Independence, the young and inexperienced Algerian government was confronted by a multitude of problems ranging from political instability to the running of a new country. This can be very well illustrated by the political struggle within the Algerian leadership that culminated by the *coup d'état* of 1965, and the establishment of a new leadership headed by President Houari Boumediene.

From the 1920s, the French were concerned with the need to house the Algerians. As a result, the idea of *Grands Ensembles* (large projects) was born and housing projects started to appear in Algiers as well as other cities in Algeria (Çelik, 1997, p. 131) (figure 3.9). Algiers and its region became important from 1959 when the Grand Alger idea was introduced and a new administrative entity to manage the metropolis was established (*GGA*, nd, p. 8). The 1966 population census revealed that Algiers had a population of 940,000. About 500,000 people were living in Algiers after Independence, as more than 400,000 Europeans left Algiers, while, between 1966 and 1970, more than 440,000 people moved to Algiers from neighbouring villages, and as a result Algiers quickly became saturated (*GGA*, nd, p. 7).

It was thought that the housing left by the French would be sufficient to meet the Algerian population's needs. This was not the case as a large part continued to live in derelict housing for many years. In the whole of Algeria, more than 700,000 housing units changed hands. While decent housing continued to be a major need, the newly established government was giving priority to economic development. Housing was not a priority for almost a decade and represented only 3.7 per cent and 5.4 per cent of public investment during the first and second development plans respectively (Chaline, 1990, p. 93).

At that time, away from the political scene, the Casbah of Algiers was the embodiment of Algerian cultural and local identity, and therefore needed special attention (Çelik, 1997, p. 185). Deluz (1988) argues that postcolonial Algiers experienced two phases of urban development. The first phase lasted until 1968 and was charac-

Figure 3.9. View toward the bay of Algiers and the Casbah.

terized by a straightforward adoption of colonial plans and projects, namely the Constantine Plan, while the second phase, from 1968 until the middle of the 1970s, concentrated on construction of new buildings and the development of a master plan for Algiers.

Algiers's outskirts were transformed by the number of new buildings whose architecture was influenced by Le Corbusier's ideas. During this period, architects such as Zehrfuss, Miquel and Pouillon established a new chapter in the urban history of Algiers by creating new building typologies, namely housing complexes such as Diar El Mahcoul, Diar Es Saada and Climat de France (figure 3.10).

The Socialist Period: the 1970s

The plans devised in the late 1960s aimed at stopping the rural exodus, and the creation of peripheral settlements around Algiers in order to establish the city as the capital of the country with all government functions (Deluz, 1988; Çelik, 1997).

This period was better known as the country's socialist era which witnessed many projects such as the Socialist Villages, large industrial projects, important prestigious facilities or urban landmarks (universities, hospitals, government institutions), and housing through the ZHUN (*Zone d'Habitat Urbaine Nouvelle* or new urban housing areas).

The country also implemented the idea of industrialization by building large-scale factories and by introducing prefabrication to the building industry, particularly for the production of housing. Housing continued to be the main source of concern and social malaise. Not only because of its shortage but also its inadequacy. In addition, other problems were added to the daily struggle of most of Algiers's population, namely shortage of water, unemployment, poor services, and the strengthening of religious groups (Çelik, 1997).

During the 1970s, Algiers started to become a metropolis, particularly after the building of important facilities such as the Palais des Nations, Hôtel El Aurassi, and the Olympic complex (figure 3.11).

Famous architects such as Oscar Niemeyer,

Figure 3.10. View of Diar El Mahcoul, Algiers.

Figure 3.11. Hôtel El Aurassi, Algiers.

Kenzo Tange, Louis Skidmore and Fernand Pouillon were all invited to design universities, and Ricardo Bofill was hired to design new housing projects. In addition, there were cooperation agreements between Algeria and other states, such as the one that led to the construction of La Foire d'Alger (exhibition hall), which was built with the help of the Chinese government.

Oscar Niemeyer probably had the most impact on Algiers urban growth as he was involved in the building of many large-scale projects such as universities and institutes, namely University of Sciences and Technology (USTA) in Bab Ezzouar, and the Faculty of Law in Ben Aknoun. Considerable emphasis was given to educational buildings, particularly higher education, thus strengthening the capital's position as a growing metropolis with services of international standing. His development scenarios for Algiers were followed by the COMEDOR (*Comité Permanent d'Etude et d'Organisation du Grand Alger*) plan of 1975, which ensured continuous urbanization of Algiers's peripheries. However, these were abandoned after the COMEDOR was dissolved in 1979.

Algiers became the scene of important national and international events and gatherings. At the same time there were many unfortunate experiences, particularly those concerned with turnkey projects that were faced with considerable technical and financial problems. In addition, the reliance on foreign architects and construction expertise and labour meant that risks were great and results were generally unsuccessful. There were very few Algerian architects and as a result the Algerian government relied on foreign experts who came from many different backgrounds and often were inexperienced and unfamiliar with Algerian realities (Deluz, 1988, pp. 134–137).

In terms of planning tools, the *Comité Permanent d'Etude et d'Organisation du Grand Alger* (COMEDOR) was responsible for Algiers's urban growth planning and control from 1968 to 1979. This institution played a crucial role in guiding and controlling Algiers's urban growth, and urban policies and practices. It was initially set up to produce a plan for the development and planning of Algiers: a *Plan d'Organisation Général* (POG) on Algiers's urban structures achievable by

2000. By the end of 1975 the POG was completed, and was adopted by decree in February 1976. This plan was the first attempt to create a major urban project for Algiers in order to strengthen its image as a capital and a metropolis.

In 1979, the *Centre National d'Etudes et de Réalisations d' Urbanisme* (CNERU) replaced the COMEDOR and was asked to develop yet another plan for Algiers: the *Plan d'Urbanisme Directeur* (PUD). By this time, the POG was being shelved leaving the urban planning scene to the PUD developed by CNERU between 1981 and 1983, and revised in 1986.

As a result, from 1968 to 1988, Algiers experienced very rapid urban growth that resulted in the construction of prominent buildings and complexes. At the same time, the planning process of Algiers was confronted by counter-productive decision-making and successive plans, mainly guided by the interests of local and regional governments, and ministries.

By the end of the 1970s, Algiers's colonial urban fabric was becoming overcrowded and deteriorated, which led to the spreading out of its periphery and the growth of rural areas such as Mitidja and Sahel.

On the political scene, the death of President Houari Boumediene in 1978 brought yet another challenge to the leadership after 13 years of stability and continuous economic and social development.

The Transition Period: the 1980s

This decade was particularly critical because of economic problems and social malaise. By the mid-1980s, there was a sharp drop in oil revenues that immediately affected the purchasing power of the majority of the population, reduced employment opportunities, and led to dramatic riots in October 1988.

The development of the *Plan d'Urbanisme Directeur* (PUD) took much of this decade. During the first 5 years, a number of important projects were started: namely, many ZHUN areas, highways and roads, renovation of Hamma area, the building of Riadh El Feth (War memorial and shopping centre), and the allocation of plots for self-building. In the early 1980s considerable effort was directed towards the containment of the city's urban growth beyond its existing boundary, the introduction of urban renovation, the improvement of road infrastructure and the eradication of the *bidonvilles* surrounding the city. Most of these goals were achieved by the middle of the decade (Deluz, 1988, p. 115).

The building of the Riadh El Feth was one of the biggest urban renewal projects in Algeria and was completed in 1984 to celebrate the thirtieth anniversary of the beginning of the War of Liberation. The monument was designed by a Polish sculptor and located near a ravine. The complex has a shopping centre and the Martyrs' Museum (figure 3.12). The Hamma renovation project was the first international architectural competition in Algeria, and was initiated in 1984 once the site was ready. This site had an area of 18 hectares and the programme consisted of the design of government facilities, housing, a national library, conference centre, and a hotel. The winning project was designed by a Canadian team, Lavalin, which was also involved in the design and building of Riadh El Feth (Deluz, 1988, p. 145) (figure 3.13).

In 1987, the *Office d'Aménagement et de Restructuration de la Zone du Hamma* (Office of Management and Restructuring of the Hamma Area) was created to oversee the implementation of the renovation project, and particularly the construction of projects such as the National Library, Hôtel Sofitel, and the high-rise housing.

It is unfortunate that most of Algiers's urban

Figure 3.12. Riad El Feth and the Martyrs' Memorial.

Figure 3.13. Hamma high-rise housing.

growth took place on agricultural land on the peripheries. Little attention was given to the rehabilitation of the Casbah and the colonial urban fabric. The only attempt was that of the renovation of the Hamma district.

The complex decision procedures and processes associated with the approval of the PUD by the parties concerned meant that its implementation could only complicate and slow down the development of the city. This was the view of the government which decided to ignore the plan and instead allowed Algiers to develop in a piecemeal fashion thus causing anarchical developments in all directions. This chaotic urbanization resulted in the creation of expensive, heterogeneous urban settlements and the consumption of considerable areas of land (GGA, nd, p. 24).

Ait-Cherkit (2000) presents a very critical view

of modern urban planning in Algeria, which saw the introduction of ZHUN as part of the PUD. He argues that these new zones, conceived as independent housing areas, had to meet a maximum density of 50 apartments per hectare. This resulted in the production of inflexible plan typologies and block configurations, which could only be cheaply built using industrialized construction methods. This, of course, resulted in monotonous housing environments without furnished open areas or adequate car parking spaces. Climatic and cultural requirements were not considered thus leading to the production of even more inadequate and inappropriate housing. Further, given that only limited public facilities were provided, the housing estates were completely deserted during the day. This situation was worsened when people that once resided in a *bidonville* or derelict area were relocated to these estates, leading to social isolation and exclusion, and the spread of delinquency.

The need for more housing pushed the government to consider mass production in order to meet the target of 100,000 units a year by the end of the 1980s. This saw the proliferation of many inadequate industrialized construction systems that achieved the quantitative requirements, but produced expensive and inadequate living environments. This figure was achieved only a few times due to many problems associated with the building technology used, namely industrialized systems.

As a result, the government decided to encourage self-building in order to increase housing numbers. Communes were involved in the subdivision of land available for urbanization, and the building of the basic infrastructure. In parallel, considerable land speculation took place and private agricultural land was subdivided and sold illegally leading to the building of illicit housing. Most of these new developments, whether legal or not, lacked basic infrastructure, public amenities and open spaces, and where infrastructure existed, it did not have the necessary maintenance or improvement.

Unfortunately, this is a pattern across Algerian cities due to lack of funding, mismanagement, and ineffectiveness of some sectors within local public institutions.

The Free Market Period: the 1990s

This was a very disturbing decade in Algerian history, dominated by the rise of terrorism and political instability. As a result, there were no noticeable urban renewal projects within Algiers apart from localized small-scale development. However, by the end of this decade, Algiers was given a new status to improve its image and to guide its urban and economic growth effectively.

The *Plan d'Urbanisme Directeur* (PUD), which took years to develop, was never adopted. In December 1990, the *Wilaya* of Algiers and the ministry in charge of urban planning (*Ministère de l'Equipement*), asked the CNERU to develop another plan for Algiers, namely the *Plan Directeur d'Aménagement et d'Urbanisme* (PDAU). They recommended that the PDAU should build on the findings and guidelines of the PUD, and should update it taking into consideration the new political and administrative changes. The PDAU has so far experienced two elaboration phases, first from 1991 to 1993, and then from 1993 to 1995. It is currently the only urban development tool available. The PUD was based on the French urban legislation of the 1960s. It did not take into consideration urban development as an evolving process influenced by other intangible factors, and assumed that most if not all urban interventions are the responsibility of the government (Ait-Cherkit, 2000*a*, p. 21).

The PDAU introduced new measures such as respect for urban economies, ecological balance, and the protection of natural and archaeological sites. It also stated that size and location of new construction should take into account existing buildings. Further, the plan stipulated that the State is not the only actor involved in the creation of new urban environments, and also identified existing urban fabric in need of urban renewal.

A considerable amount of Algiers development had been informal and uncontrolled, which resulted in a shocking growth of the urban area: from 7,500 ha in 1970 to 25,000 ha in 1990 (GGA, nd, p. iv). Many large formal and informal areas have grown on the peripheries including the new ZHUN. In addition, most of the old areas in Algiers, such as Bab El Oued, Belouizdad, Al Madania and the Casbah are in a poor state. These neighbourhoods are restricting the social development of the population and encouraging the spread of poverty.

The Casbah, which is overpopulated and in need of urgent rehabilitation (figure 3.14), presents an agonizing problem. Once a symbol of resistance to occupation and later a place of art and responsive architecture and planning, it has now fallen into a state of neglect and attracted large numbers of temporary informal settlers whose only aim is to obtain decent accommodation elsewhere. It is a national historic monument that, in 1990, was classified by UNESCO as a World Heritage site, therefore putting pressure on the Algerian authorities to restore it and avoid its destruction. As a result, a number of attempts have been made to rehabilitate the site to its former magnificence.

Algiers, with its decaying and inadequate built environment and its anarchic urban development, contributed to the increase of social discontent and spatial segregation amongst its population, a situation worsened by the lack of effective urban planning and development control.

Due to these factors and the political instability at the time, it was difficult for the city to change from a disorganized and unplanned metropolis into an international hub at the gate of Africa. It was claimed that Algiers should establish an urban growth plan that would encourage expan-

Figure 3.14. Dereliction in the Casbah.

sion and development at the metropolitan scale and not at the local (*commune*) scale as had been happening. This growth had resulted in a clash of interests between the local authorities and a lack of funding, in addition to unprecedented mismanagement, speculation and corruption. The idea was for Algiers to attract national and international investment and become competitive during this era of tough free market competition. In this respect, Algiers needed to improve its urban infrastructure and public amenities, resolve the access to national and regional institutions located in the city centre, and to start considering the bay as an area for future development for public and commercial buildings.

In order to implement the new planning strategies, a new institution, the *Gouvernorat d'Alger*, was created in 1997, and the idea to re-conquer the centre of Algiers was born. The new project was called the *Grand Projet Urbain* (Large Urban Project). This consisted of the introduction of three aspects:

- a development scheme that would protect land use appropriation and the creation of new urban centers to balance the metropolitan city;
- the establishment of a new agency, *Agence d'Aménagement et d'Urbanisme* (Agency for Planning and Urbanism), to ensure a harmonious and lasting urban development strategy;
- the use of Geographic Information Systems (GIS) to improve management and decision-making and obtain up-to-date information.

Planning strategies of the 1990s proposed the establishment of a green belt around the city to control its urban extension and offer open space for recreation and sport.

In fact, the *Gouvernorat* of Algiers established a new urban planning policy, which centred around six objectives:

- the *Grand Projet Urbain*, with emphasis on economic development;
- valorization of neighbourhoods;
- re-launch of the urban economy;
- strengthening of road and other infrastructure;
- preservation and protection of the environment;
- modernization of public institutions. (GGA, nd, p. 47; Hammache, 2000*b*, p. 30)

It was clear that Algiers felt the need for a substantial change in its planning and development in order to reach the level of a regional metropolis. It is interesting to note that the city played a more international role in the 1970s than it did in the 1980s and 1990s, and this despite the globalization effect that many cities in the Mediterranean and Africa had experienced during the last three decades. The internationalization of the economy and the introduction of a free market, had not affected Algiers either as a global or metropolitan city. Indeed, while Algiers could easily have become the hub or headquarters for many international companies and a venue for international and regional events, instead internationalization passed the city by.

In order to overcome this problem, the new administration attempted to strengthen the city's role by attributing to it three new functions: metropolitan functions; international functions; and valorization of cultural, scientific and technological functions.

Six major urban projects were then suggested as follows:

- the Casbah and *La Marine* quarter, the colonial urban fabric within the city centre, and the seaport were to become a central urban area with commercial emphasis. Special attention would be given to the port to integrate it into the city through new economic activities, recreational spaces and functions, transportation and fishing;
- identification of specific location(s) for a business centre(s);
- combination of existing universities with recreational activities;

- development of tourism activity and recreation centre or axis;
- improvement of the western sea front in order to strengthen its tourism attraction;
- development of a national and international business centre in the west of Algiers (Hammache, 2000*b*, p. 30) (figure 3.15).

It can be said that the emphasis of the new urban projects was economic and recreational; economic in the sense that tourism should be introduced in some areas and strengthened in others at the same time as the establishment of business centres. However, these objectives fell short of their expected outcome due to recurring political and financial complications. Algiers thus remained an incomplete metropolis like many other cities of the developing world. These cities have experienced rapid and largely uncontrolled urbanization, and are unable to offer the necessary public amenities due to lack of financial resources and poor organizational planning.

Algiers in the New Millennium

Algiers, apart from spreading beyond its established urban perimeter, has shown a considerable delay in meeting its population's needs in terms of public services such as transportation and housing supply. Current estimates show that the city has a population of around 2,600,000 distributed within fifty-seven communes, and has experienced a population increase of about half a million every decade.

The *Gouvernorat d'Alger* aimed at turning Algiers into an international metropolis by

METROPOLITAN ALGIERS IN 2000

AFTER: INCT (1999) & EPAU & SIAAL (2000)

Figure 3.15. Algiers central and metropolitan areas.

strengthening its role regionally and internationally. This was particularly encouraging given that Algiers's geo-strategic location should enable it to become an important destination linking the Arab world and Africa to Europe and the Mediterranean region. Why this has not been achieved is a complex matter that can be linked to several reasons, which are discussed below.

Algiers's growth has been the result of a combination of at least three factors:

- a dramatic increase in its population. Table 1 below shows the increase in population over the last four decades;
- increased economic activity particularly during the 1960s and 1970s, which saw the building of important industrial installations; and
- Algiers's status as capital of the country, which meant that all the institutional and government facilities are located within the city, therefore requiring more infrastructure (Hammache, 2000*a*, p. 12).

The *Gouvernorat d'Alger* was suspended in 2000 by the new President of the Republic Abdelaziz Bouteflika, who claimed that its creation was unconstitutional, since it was functioning like a ministry, with the Governor acting as a minister for Algiers. As a result, the *Wilaya* of Algiers took over as the sole administrative and political decision-maker. These consecutive changes in management and decision-making have been counter-productive and only slowed down the development of the city and its establishment as an important metropolis in the area.

The EPAU & SIAAL study (2000) concluded that the *Grands Projets* initiative would not be successful unless Algiers's urban fabric evolved. The study compared Algiers with a number of European cities in order to develop some performance criteria. It concluded by setting out a number of recommendations at both regional and urban levels. At the regional level, the emphasis was on environmental planning and land registry and control. At the urban level, the study stressed the importance of improving economic activity, strengthening the image of the city, improving the existing urban fabric and promoting tourism.

Algiers and its metropolitan region have grown incoherently around a potentially efficient vehicular network. This network, which seems to work at the city level, has never been effectively connected at the neighbourhood level. This is due to considerable urban development characterized by the fact that ZHUN estates and spontaneous and anarchic constructions are all disconnected and inadequately laid out. As a result, a considerable amount of land is wasted, and the road network is fragmented and inefficient.

The current decrease in interest in public health in Algiers, illustrated by the increasing lack of cleanliness (little rubbish collection), clearly shows institutional disorganization, compounded by complex social and cultural transformations. In addition, the state of decay of the colonial built environment has been exacerbated by overcrowding and its inability to accommodate this (De Ameyar, 2002).

Table 1.1. Population growth within Algiers.

Designation	*RGPH 1966*	*RGPH 1977*	*RGPH 1987*	*RGPH 1998*
Algiers	10 communes	13 communes	28 communes	28 communes
Inhabitants	943,551	1,353,826	1,507,241	1,585,609
Gouvernorat d'Alger (57 communes)	1,094,851	1,641,521	2,015,374	2,562,428

(*Source:* ONS, 1992, 1998)

It was hoped that the globalization of the economy, and the increased flow of ideas and information and, to a lesser degree in the case of Algeria, investment, would help improve conditions in the city. Unfortunately, there are still considerable administrative and political hurdles and complications blocking or discouraging foreign capital investment.

The earthquake that shook the Algiers area in May 2003 has revived the discourse on the responsibilities of the local authorities and the purpose of building controls, as it appeared that old buildings withstood the earthquake much better than the new ones. New housing built by cooperatives and individual builders were reduced to rubble. The architectural profession has also been put to the test and criticized for failing to address critical issues facing Algerian cities, for producing inadequate building forms, and for the lack of involvement in the improvement of building quality and safety (Boudaoud, 2003).

This is yet another challenge to be added to the complex list of problems facing Algiers's development and growth control. It is a wake up call for the Algerian government, the local authorities, and the urban and architectural professions. It is time for Algiers to compare itself with successful African and Arab cities that were able to establish themselves as regional hubs by attracting businesses and ensuring a sustainable economic and social development.

At the metropolitan level, there is an urgent need for more adequate housing, public amenities, public transportation such as the Metro, and even a new airport. In addition, the Casbah renovation project should be addressed effectively and quickly implemented, otherwise the continuous decay and abuse that is taking place in this rare urban heritage will only speed up its disappearance.

The *Gouvernorat d'Alger* proposed a number of measures for the rescue of the Casbah as part of its *Grand Projet Urbain*:

- reduction of the population density by moving out at least 5,000 inhabitants;
- preservation of existing open spaces;
- introduction of local urban businesses;
- promotion of the rehabilitation of arts and crafts activities and businesses that should benefit the Casbah and ensure its sustainability;
- development of a policy that will provide opportunities for employment and vocational training to the youth of the area;
- provision of the means for delinquents to reintegrate into normal life.

It is therefore essential that the Casbah of Algiers be rehabilitated both physically and socially. The mere renovation of its physical condition will not improve its reputation and attract national and international visitors and investors unless the social and cultural problems are also tackled and resolved in full.

In summary, Algiers has experienced an unbalanced urban growth most of which occurred during the boom years in the 1970s and 1980s. Prior to this the Algerian government was still influenced by the 'French project' for Algiers, and after, the socio-economic and political situation forced its urban development and renewal projects into a secondary position. The recent changes in Algiers's political and administrative status appear to be counter-productive and will only ensure its unplanned and uncoordinated growth. They will also prevent the city from becoming a regional and international economic and tourist destination. Having said that, the dilemma regarding Algiers globalizing or not remains on the agenda of national and local governments, and the private sector. The recent changes to the city's status appear however to show that the government has little intention of forcing the capital into an international hub just yet. In the meantime, there are many hurdles to be met, such as the renovation of the Casbah, meeting the housing shortage, the building of a new airport, and improving the city's infrastructure and image.

Algiers's urban development policies and strategies should be reviewed inline with the current priorities. Less emphasis should be given to the development of more planning tools, because many existing studies are available and valid. More emphasis should be given to their implementation and to an active participation of national and international developers.

Conclusion

This chapter has examined Algiers's spatial structure and studied its urban transformation during the colonial and postcolonial periods, by concentrating on important phases of its growth and their impact on social and economic development on the one hand, and the perceived image of the city on the other.

The first part of the chapter focused on the colonial spatial development of Algiers and demonstrated that the initial French intervention was founded on a two-part conviction: the superiority of Western civilization and the inferiority of the indigenous civilization. This conviction justified the imposition of 'modern' European planning, based upon regularity and large perspectives, onto a spatial organization considered chaotic and anarchic. Thus, with colonization, a spatial upheaval occurred leading to the imposition of the European city on the old precolonial structures, which gradually suffocated under alien planning and architectural concepts.

As the colonial capital of Algeria, Algiers underwent the most profound transformation in its basic functions and structures. Within this scope Algiers's colonial urban history saw five dominant periods, during which it evolved from a medium-sized North African *medina*, to an African metropolis.

The analysis above shows that the first decade of colonial rule in Algiers was a period of uncertainty. The political status of colonial Algeria was not clearly defined and the French were still wondering whether to settle. The French did not originally consider the possibility of creating a new urban structure next to the pre-existing urban fabric, which was typical of French colonization in other countries. The destruction of various parts of the Casbah, when vast sites outside the walls existed, upon which it would have been easy to construct, reflected the military character of the intervention. The first spatial modifications took place in the lower, relatively flat, precolonial city. Hence, a Place d'Armes was created to accommodate troop gatherings and parades. Many streets were straightened and widened and various buildings were transformed and adapted for new needs. For a long time, military zones were spatial hindrances to the development of the city. No civil construction was allowed outside the walled city and the army monopolized the flat sites. The result was an incoherent policy leading to a continual conflict between the concept of Algiers as a military centre and the idea of developing the city into a commercial emporium.

The transformation of the conquest into colonization attracted new immigrants who settled in the 'Westernized' part of the Casbah, almost depopulated of its Moslem citizens. New living spaces were urgently needed. A first spatial expansion of the city occurred with the construction of new ramparts and a new layout of streets characterized by large arteries connecting the Casbah and the newly developed European quarter. The arcaded Boulevard de la République was incontestably the apex of the large axes, conceived as a continuous perspective. With its monumental façades, it became a symbol of colonial Algiers for the people coming by sea from France, reassuring them of being home. In fact, it was only an architectural screen hiding uncontrolled development behind.

The linking of the two urban fabrics required the opening up of new streets and squares inside the old core. Military institutions dominated the flat sites within the intramural city. Thus, the available land for city expansion was limited, forcing the civil authorities gradually to destroy the old city in order to rebuild it to meet their new needs. It resulted in contiguity of the two types of space. This confrontation though carried out through violent processes, by destruction and population transfers, could not overcome the original Casbah and the French, therefore, were content to conserve, but dominate it.

Three main features characterized the urban development of Algiers between 1880 and 1914. Firstly, the success of viticulture in Algeria generated considerable wealth. Funds, insufficient during the preceding decades, became abundant giving an impulse to the process of urbanization in the whole colony. Secondly, many military sites were given to the city allowing its expansion southwards. Thirdly, there was the tendency towards the gradual displacement of the urban centre, from its original location, southwards to the new European quarters. Gradually, the Casbah of Algiers was emptied of its original functions and then of its position as an urban centre, becoming a marginal quarter, dilapidated and crowded with an indigenous proletarian population. An important number of the wealthy Algerians abandoned the Casbah to live in new European quarters.

In contrast to what happened in French cities in the late eighteenth century, where sites available after the destruction of the medieval fortifications were used to create a peaceful area between the old nucleus and the new urban development, those of Algiers became building-yards invaded by speculators and constructors.

The period between the two World Wars can be best summarized as the 'ruralization' of cities. Rural and nomadic people were squeezed off their land and flocked to the city of Algiers where no provision had been made for them. The Casbah, which traditionally represented a balanced social and economic *milieu*, tended to become a social ghetto for proletarian and rural people. In the 1930s, Algiers saw a rapid and extensive increase in the Moslem population, worsening the process of urbanization. The high population growth created a drastic housing shortage. Migrants from the countryside who could not find living spaces inside the Casbah, started to construct their own shelters on unbuilt areas, mostly on sloping sites. These spontaneous houses, which did not follow either the Casbah's pattern or the European model, formed *bidonvilles*, which gradually ringed the city.

In short, Algiers 'dilated' from a single core, the Casbah, to a complex conglomerate of urban units, stretching out in a long curve along the coastline.

The second part of this chapter discussed postcolonial urban transformations of Algiers, and revealed that the city needs to strengthen its urban development policies and implement them effectively. It is believed that the planning tools are available, but need to be used with the right instruments to ensure that the city grows under control by tackling existing urban issues. There are many urban challenges facing Algiers particularly the safeguard of the Casbah.

In order for Algiers to achieve its goal of becoming a regional metropolis it must:

- decide on an efficient urban planning and management tool, which should have as a priority the establishment of Algiers as a regional metropolis in the western Mediterranean region;

- produce a heritage conservation policy that would ensure the rehabilitation of the Casbah which is a World Heritage site in need of urgent intervention;
- carry out urban renewal within the colonial neighbourhoods that are becoming derelict;
- improve the city's transportation facilities (road infrastructure, public transportation, seaport and airport facilities);
- complete the new airport to allow an increase in numbers of visitors and to offer more comfortable facilities;
- improve the financial institutions by the strengthening of the banking system and facilities;
- encourage foreign investment and joint ventures, particularly to re-establish the tourism industry by building, for instance hotels, and leisure centres;
- promote the city in regional and international events by offering its services (financial, administrative, transportation, cultural, tourism);
- produce an environmental protection act with the necessary implementation measures and means that would protect the environment and public health.

In addition, educational aspects related to sustainability, cleanliness and hygiene, and anti-social behaviour and crime, targeting public and private properties as well as citizens should be addressed.

Globalization has not affected Algiers in the way it did in other African and Arab conurbations. The city has retained its 'corner shop' culture and has not opened fully its doors to multi-nationals and the movement of capital. Further, tourism has never really blossomed in either Algiers or the rest of Algeria for many reasons, which has greatly contributed to the decline of the Casbah and the poor financial facilities.

While the city continues to grow as a group of largely independent neighbourhoods rather than as a single entity, the lack of reliable information about its urbanization is one of the major causes for concern. This is primarily because scarcity of funds and political instability have led to a shortage of resources that should be allocated for research and consulting studies.

From *Gouvernorat d'Alger* to the *Wilaya* of Algiers, it is hoped that the new administration will attempt to strengthen the city's image and ensure that the capital will regain its regional and international reputation, and become an attractive destination for investors and visitors alike.

REFERENCES

Abu-Lughod, J. (1976) Development in North African urbanism: the process of decolonization, in Berry, B. (ed.) *Urbanization and Counter-Urbanization*. London: Sage, pp. 191–212.

Ait-Cherkit, S. (2000) Les instruments d'urbanisme traditionnels, in EPAU & SIAAL, *Alger Métropole: Région – Ville – Quartier*. Ecole Polytechnique d'Architecture et d'Urbanisme (EPAU), Alger, in cooperation with and published by Städtebau-Institut, Stuttgart University (SIAAL), Germany, and Deutsche Gesellschaft für Technische Zusammenarbeit GmbH (GTZ), Eschborn, Germany, pp. 21–22.

Beguin, F. *et al.* (1983) *Arabisances*. Paris: Bordas.

Boudaoud, A. (2003) L'architecture: une profession clochardisée. Interview by K. Abdelkamel. *Liberté Algérie*, Dossier edition, 25 May.

Çelik, Z. (1997) *Urban Forms and Colonial Confrontations: Algiers under French rule*. Berkeley and Los Angeles: University of California Press.

Chaline, C. (1990) *Les villes du monde arabe*. Masson: Paris.

Chennaoui, Y. (2000*a*) Les processus de structuration du territoire algérois, in EPAU & SIAAL, *Alger Métropole: Région – Ville – Quartier*. Ecole Polytechnique d'Architecture et d'Urbanisme (EPAU), Alger in co-operation with Städtebau-Institut, Stuttgart University (SIAAL), Germany, and Deutsche Gesellschaft für Technische Zusammenarbeit GmbH (GTZ), Eschborn, Germany.

Chennaoui, Y. (2000*b*) Alger: portrait historique, in EPAU & SIAAL (2000) in EPAU & SIAAL, *Alger Métropole: Région – Ville – Quartier*. Ecole Polytechnique d'Architecture et d'Urbanisme (EPAU), Alger in cooperation with Städtebau-Institut, Stuttgart University (SIAAL), Germany, and Deutsche Gesellschaft für Technische Zusammenarbeit GmbH (GTZ), Eschborn, Germany, pp. 15–17.

Comedor (1972) *Etude pour la rénovation et la construction de la Casbah.* Algiers: Atelier Casbah.

Curtis, W. (1986) *Le Corbusier: Ideas and forms.* Oxford: Phaidon.

Davis, A. (ed.) (1983) Tunisian arcades. *Middle East Construction,* January, p. 15.

De Ameyar, H. (2002) *Alger: du gouvernorat a la wilaya. Quel statut pour la capitale?* Special Report, *Liberté Algérie,* 2 December, Algiers.

Deluz, J. (1980) Alger 1962: L'héritage. *Technique & Architecture,* No. 329, pp. 41–46.

Deluz, J. (1988) *L'urbanisme et l'architecture d'Alger.* Algiers: Office des Publications Universitaires: Alger.

Emery, P. (1980) L'architecture en Algérie, 1930–1962. *Technique & Architecture,* No. 329, pp. 54–57.

EPAU & SIAAL (2000) *Alger Métropole: Région – Ville – Quartier.* Ecole Polytechnique d'Architecture et d'Urban isme (EPAU), Alger in cooperation with Städtebau-Institut, Stuttgart University (SIAAL), Germany, and Deutsche Gesellschaft für Technische Zusammenarbeit GmbH (GTZ), Eschborn, Germany.

GGA (Gouvernorat du Grand Alger) (nd) *Alger capitale du 21eme siècle: Le Grand Projet Urbain.* Algiers: Gouvernorat du Grand Alger.

Hammache, S. (2000*a*) Contraintes et potentialités, in EPAU & SIAAL, *Alger Métropole: Région – Ville – Quartier.* Ecole Polytechnique d'Architecture et d'Urbanisme (EPAU), Alger in cooperation with Städtebau-Institut, Stuttgart University (SIAAL), Germany, and Deutsche Gesellschaft für Technische Zusammenarbeit GmbH (GTZ), Eschborn, Germany, pp. 12–14.

Hammache, S. (2000*b*) Les Grands Projets Urbains en cours, in EPAU & SIAAL, *Alger Métropole: Région – Ville – Quartier.* Ecole Polytechnique d'Architecture et d'Urbanisme (EPAU), Alger in cooperation with Städtebau-Institut, Stuttgart University (SIAAL), Germany, and Deutsche Gesellschaft für Technische Zusammenarbeit GmbH (GTZ), Eschborn, Germany, pp. 30–31.

Institut Français d'Architecture (1984) *Alger.* Paris: Bulletin d'information architecturale.

Institut National de Cartographie et de Télédétection (1999) *Alger: plan des rues.* Algiers: INCT.

Kecir, H. (1980) *D'icosium à la Casbah: processus du développement et de la construction de la ville d'Alger.* Paris: University Paris VIII.

Laffont, P. (1981) *L'Algérie des français.* Paris: Bordas.

Lawless, R. (1979) Algeria: transformation of a colonial system under a planned economy, in Obudho, R. and El-Shaks, S. (eds.) *Development of Urban Systems in Africa.* New York: Praeger, pp. 79–98.

Lespes, R. (1930) *Alger: étude de géographie et d'histoire urbaine.* Paris: Felix Alcan.

Miege, J. (1985) Algiers: colonial metropolis (1830–1961), in Ross, R. and Telkamp, G. (eds.) *Colonial Cities.* Dordrecht : Martinus Nijhoff, pp. 171–179.

Miquel, L. (1980) Alger, Le Corbusier et le Groupe CIAM-ALGER. *Technique & Architecture,* No. 329, pp. 58–59.

Mutin, G. (1982) *La politique urbaine en Algérie, politiques urbaines dans le monde arabe.* Lyon: C.N.R.S.

ONS (Office National des Statistiques) (1992) Evolution des agglomérations 1966–1977–1987. *Collections Statistiques,* no. 38. Algiers: ONS.

ONS (Office National des Statistiques) (1998) RGPH 1998 Résultats Préliminaires, *ONS Bulletin,* no. 270. Algiers: ONS.

Osmani, M. (1988) Physical Form and Spatial Structure of Colonial Cities in Algeria. Unpublished MPhil Thesis, Joint Centre for Urban Design, Oxford Brookes University, Oxford.

Renaudot, F. (1979) *L'histoire des Français en Algérie: 1830–1962.* Paris: Laffont.

Royer, J. (1932) *Urbanisme aux colonies et dans les pays tropicaux.* La Charite sur Loire: Communications & Rapports du Congrès International de l'Urbanisme aux Colonies et dans les Pays de Latitude.

Stewart, D.J. (2001) Middle East urban studies: identity and meaning. *Urban Geography,* **22**(2), pp. 175–181.

Chapter 4

Globalization and the Search for Modern Local Architecture: Learning from Baghdad

Hoshiar Nooraddin

I visited Baghdad following the last Gulf War, which ended with a change in the ruling power in April 2003. A city which had taught me architecture and love had taken on a new shape and a new reality. I saw destroyed or burned buildings, streets with no real life, a society trying to answer endless questions: why all this and how to rise again? But I had no doubt that a new era with a flourishing society will emerge for this great nation.

Introduction

Globalization is the most recent chapter in the annals of dominance whose origins are to be found in the earliest empires – Babylon, Egypt and Greece – and whose progress may be charted through the history of the world's major powers – the Islamic Empire, British Empire and Soviet Union to today's major industrial powers – the United States and European Union. Historically complex, globalization has been applied, developed and justified under different names, reasons and means, and has left its mark on the architecture of all the societies it has touched.

There are many definitions of globalization, reflecting the varied assessments of the phenomenon. For example, it has been referred to as a process of economic liberalization and free market creation where individuals, corporations and nation-states can reach around the world. Another definition formulates the process as a fundamental change in the spatial-temporal contours of social existence that contributes to a change in the human experience of space and territory.

On the other hand, globalization is criticized because it is dominated by major economic powers and according to their interests. It is said to limit the power of nation-states to protect their domestic interests and that it threatens local culture, particularly in developing countries. It has been criticized for neglecting the diversity of our international societies and for challenging local control, local national interests, and local resources (Stiglitz, 2002, pp. 3–8).

International architectural movements have established a global design philosophy, which has facilitated the practice of modern architecture beyond traditional national borders. Thus, the contemporary architecture of almost every nation has developed over time from the efforts of architects belonging to different nations and with different architectural styles and frameworks.

Globalization has provided architecture with advanced technologies and international business services that enable a new way of working within an international framework. The driving forces are the Internet, digitization, and high speed data networks that connect architects around the world. These technical facilities make information exchange and the creation of detailed architectural drawings possible from anywhere in the world. Because this process is based on advancing the global economy and is realized through the international market, the globalization of the architectural practice is sensitive to economic conditions (Engardio *et al.*, 2003).

Globalization and the Architecture of Baghdad

Baghdad is located on the both banks of the river Tigris in the central area of Iraq. With a current population of about five million, it is the country's economic, political, communication, cultural, and financial centre.

Iraq is one of the oldest civilizations and is often referred to as the 'cradle of civilization'. Throughout its history it has seen the playing out of the processes of dominance and domination. The interplay of war and peace has been an important characteristic of this rich nation.

In more recent times colonialism has been an important chapter in the country's history because of the role it played in establishing the new state of Iraq and in introducing Western culture into the country. It contributed also to the opening up of the local Iraqi market to international firms, particularly the British. Building and construction were among activities where international companies became involved at an early stage, especially after the First World War, and with almost no local debate or participation in design.

Local architects did not become involved in design and development in the country until the 1930s when a small group of young Iraqi architects returned following training overseas, mostly in England. They played an important role in increasing the influence of the modern Western architecture in Iraq. However, by the 1950s increasing national movements in Iraq were encouraging local architects to search for a modern national architectural identity. By the 1960s Iraqi art and architecture had developed a distinct indigenous style and local movements concerned with different aspects of modern Iraqi design had been established.

During the 1970s and 1980s, Baghdad experienced major construction activity as a direct consequence of the economic boom resulting from increasing oil revenues. These projects changed dramatically the city's local traditional architecture. A large number of local and international architects belonging to different architectural movements participated in the process.

Three major regional wars since the 1980s together with international sanctions following the 1991 Gulf War caused a dramatic slow down in construction. And these events eliminated the country's ability to participate in the process of globalization. However, during its history, Iraq has shown a remarkable ability to rise, change, and develop. Today there is a growing commitment among the local architects to rebuild and develop Iraq following the last Gulf War which ended in April 2003 with the change of the ruling power. It raises hopes of developing a bright new future for the local architecture of Iraq.

Baghdad Past and Present

The building of Baghdad represents an important era in the urban history of Iraq. It was the tradition for each dominating power to build its own capital city – a tradition which began with the early civilizations of Mesopotamia. Ancient Mesopotamia (from the Greek meaning 'the land between the rivers') was the first civilization to emerge in the Near East around 5000 BC (Morris, 1994, pp. 5–8). The area is blessed with two great rivers, the Tigris and Euphrates, and adequate rainfall. Construction of water works to control the floods of the two rivers provided extremely fertile land which supported Man's first cities and civilizations. The region attracted different peoples who settled in the area and contributed to building new civilizations such as Babylon in the south and Assyria in the north. The area became an ethnic mix – a characteristic which continues in Iraq today.

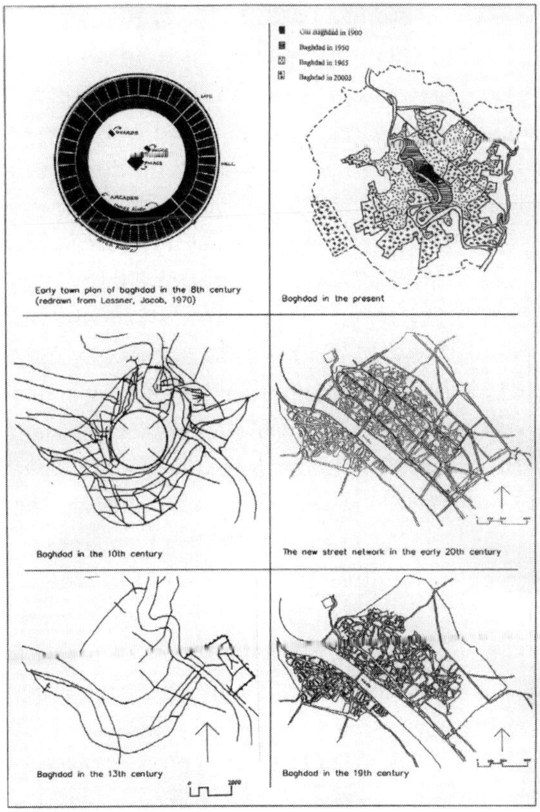

Figure 4.1. Development of Baghdad through history.

In the seventh century AD Iraq became part of the Islamic world and in the eighth, the centre of the Islamic Empire. Baghdad was founded by the Abbasids who ruled the Islamic world for five centuries (750–1258 AD).[1] Historical documents show that the city was built on the west bank of the Tigris in 145 H/762 AD (Lasner, 1970, p. 27), and therefore, compared with the ancient cities of Iraq such as Babylon, Baghdad is a young city.

The original city was circular with a radial street layout and became known as the Round City (figure 4.1). Shortly after its construction, increasing numbers of newcomers from other areas led to people settling outside the city wall to form suburbs, which extended out to the eastern bank of the Tigris.

During the next two hundred years the regular shape of the Round City was gradually transformed into an irregular and organic form and most of the original design disappeared (figure 4.1). This was probably the result of several forces such as the population growth, decline of the central power of the Abbasids, and periodic demolition and reconstruction. It also reflects the people's role in creating organic forms by occupying and building parts of the street and changing the original designs of buildings (Nooraddin, 1996, pp. 61–65).

In 836 AD the seat of power moved to Samarra following a political struggle, but returned to Baghdad again in 892 AD. The new rulers chose the eastern suburban area of Baghdad, named the Rusafa, as their centre of power. During the eleventh century, the organic urban structure of the Rusafa was surrounded by walls, forming a rectangle.

Baghdad suffered major destruction at the

hands of the Mongols in the thirteenth century, which totally changed the city's status from a flourishing power centre to a small, weak and degraded backwater. It was conquered again in 1683 by the Ottomans and became a province with a Governor (*Pasha*). While this gave the city political stability after a long period of struggle and instability, it did not regain its role as a centre of power and culture.

In 1917 the British took Baghdad as a result of the First World War and the collapse of the Ottoman Empire, which was allied to Germany. However, Iraq's first contact with the British dates back to 1763 when the English East India company set up a factory in the port of Basra to support commercial enterprises in the region. In 1802, a British Consulate opened in Baghdad, which stimulated British commercial activity in the area, such as importing British steamboats in 1836 and laying a telegraphic line in 1861.

At the end of nineteenth century, the Ottomans decided to diversify their economic relationships with European powers, particularly with the Germans. This policy resulted in a series of high profile projects realized in collaboration with Germany such as the Berlin to Baghdad railway via Istanbul and Konya, which was built in 1899. The Young Turks' revolution of 1908 also used German systems as a model to reform the Ottoman political and economic systems.[2] The British considered these events a threat to their imperial interests and were the impetus for their occupation of Iraq during the First World War.

Although economic considerations were the driving force that created the colonial era, yet it had major impacts on many other aspects of Iraqi society, which changed radically from the Ottoman period. There were changes in the administrative and legal systems, trade, education, art and lifestyle. As these changes took root there emerged new forms of architecture and city planning. The impact of Western architecture, particularly that of Britain, on Baghdad grew gradually after 1918, and two decades later modern European architecture had become the major source of design in the city.

Baghdad Emerging as the Capital City of Iraq

A direct consequence of the collapse of the Ottoman Empire and the emergence of Western colonization was the establishment of new nation-states in the Middle East, among them Iraq. The British brought together the three Ottoman provinces of Baghdad, Basra and Mosul to establish a new state and named the country Iraq under a 1920 League of Nations mandate. In 1922 a new government with a king at its head was installed; this government was in effect a British puppet. Baghdad became the capital city of the new state, the position it had lost after the collapse of the Abbasid Empire and destruction at the hands of the Mongols.

The British were basically concerned with controlling their new vassal. Therefore they developed and enforced a comprehensive treaty and alliance between Iraq and Britain. They also introduced a new constitution for the country similar to the British-style constitutional monarchy and parliament. The treaty guaranteed commercial freedom and religious tolerance, but British advisers were to control and decide foreign, military, juridical and financial matters.

In 1932 Iraq was allowed to enter the League of Nations as an independent nation. But in reality the country continued under British control indirectly, and Britain maintained considerable influence on Iraqi national affairs until the royal family and its government were overthrown in 1958 and Iraq became a republic.

The governments which followed established centrally-controlled political systems that influenced all aspects of life in the country. As a part of this the practice of architecture by local architects and the use of international architectural partners were regulated and controlled by different Iraqi legal codes. Under the terms of the free education system all architects educated in Iraq had to work for the government. Also according to the rules, the central authority controlled the use of international architectural firms.

The Emergence of the Modern Architecture of Baghdad

British colonization marked the end of the old system of administration in Iraq and the start of a new way of organizing and managing the country, which affected many aspects of Iraqi society. Amongst these was the architecture of Baghdad which became subject to new town planning rules and new building and land regulation codes. New architecture and new construction materials and technology were introduced. These changes were important for the architectural history of Baghdad and the whole nation because they opened the local architectural market to international architects; I will return to these changes below.

Changing the Legal System

Iraq was the birth place of the first system of law, which was promulgated by King Hammurabi in Babylon in the eighteenth century BC and became known as the Hammurabi Code. The Code covered production activities and the relationship between people; several paragraphs were devoted to building works and construction contracts.

Iraq was also the place where the Islamic jurisprudence (*fiqh*) was developed, which contributed to the development of Islamic law (*al-shari'ah*). This form of jurisprudence and its theory were first introduced by Imam Abu Hanifah (d. 150 H/767 AD) and later by other scholars. The method depended on using analogical deduction (*qia's*) to derive a particular judgement in any field such as the built environment or the urban society (Nooraddin, 1996, pp. 66–73).[3] In a broader sense, these early scholars attempted first to connect the ideal theory of Islam to practice and to create moral and legal obligations according to Islamic principles, and secondly to keep the practice of the legal system under control. The judgements were time-, place-, and case-related and could be changed according to changes in these factors. This system gave its users extensive access to the decision-making process by reflecting their customs and desires in shaping and organizing their built environment. This innovative and flexible way of thinking continued until the thirteenth century when Baghdad was invaded by the Moghuls. The fall of the Abbasid Empire contributed to the decline in the quality of cultural development in Iraq. As a consequence, jurisprudence lost its innovative nature and became confined to the judgements and analogical deduction methods of the established traditions. Thus in the late part of Islamic civilization, the old judgements were continuously reapplied without any updating.

In 1869, the Ottoman administration tried to modernize Islamic law through the *tanzimat* reforms by collecting and standardizing the existing rules of Islamic law (Lewis, 2001).[4] The building codes and land regulations of this reformed law system were developed in order to establish common guidelines to control and organize settlements throughout the Ottoman Empire. However, the collapse of the Ottoman Empire at the end of the First World War halted this attempt to establish a modern alternative of Islamic law.

Following the First World War, the British colonial administration and subsequent national authorities of Iraq gradually and systematically replaced the Ottoman system of law, primarily with elements of the British legal system but also from other Western systems. This was done for a number of reasons including facilitating the Iraqi market for the benefit of British global interests. Successive administrations of royal and republican governments continued adapting other international law systems to establish or develop the Iraqi legal system (Barzinji, 1978).

Following British occupation of Iraq during the First World War, the new colonial administration issued the Panel Law of Baghdad. This was valid until 1969 when the Iraqi government replaced it with Law 111, known as the Iraqi Panel Law.

However, Ottoman civil law (*Mejallat al-Ahkam al-Adliyah*) remained in force until 1953, when it was replaced by the Iraqi civil code issued in 1951 and instituted two years later. This new civil law embodied a combination of Islamic law (*al-shari'ah*) and Western civil law. The form of Western civil law was an adaptation of Egyptian civil law which, in turn, was founded on French civil law. Building regulations were included in the civil code, which laid down the relationship between the parties involved in construction activities and also contained building specifications and general contracts and conditions of work.

Changing Baghdad's Town Planning

Changes in local architecture and the urban fabric of Baghdad began slowly after the First World War and increased dramatically after the Second World War. Because Baghdad is the capital city and centre of power, it experienced urban transformation and the introduction of new architectural styles before other Iraqi cities, and subsequently influenced the introduction of modern architecture elsewhere in the country.

The introduction of modern architecture took place in parallel with the development of new town planning for Baghdad, but in different ways. Baghdad's architecture was the concern of local architects, while the city's planning was developed almost entirely by international firms which applied typical models of modern town planning; only in the later period did local planners begin to influence the planning debate.

During the Ottoman period, Baghdad did not undergo any major transformation of its layout until the reign of Midhat Pasha who, in 1869, initiated different modernization programmes following the *tanzimat* reforms, which involved the whole Ottoman Empire.

Earlier transformations of Baghdad had always consisted of additions to existing built-up areas. This trend changed early in the twentieth century when, as part of the modernization process, new streets which cut through the existing fabric were introduced. This new street system with its avenues and squares formed the basis of Baghdad's urban transformation. Further, the wide, straight thoroughfares encouraged another important element in the transformation: the new European design of the buildings along these streets. The major consequence of this process was the establishment of a different street life and a different relationship between indoor and outdoor spaces, which constituted a character totally different from the traditional urban fabric. Except for some old areas of Baghdad, the present architecture of the city belongs to different twentieth-century architectural movements.

In 1914, the Ottoman administration started with the opening of Al-Rasheed Street as a modern avenue to facilitate transportation to and within the city and to develop commercial trade in the central area (figure 4.1). The street was regular, straight and wide, its central part designed for carriages, and the sidewalks, shaded by arcades, for pedestrians. The traditional narrow and winding street network of the old city fabric was unsuitable for carriages or other vehicular traffic. Al-Rasheed Street extended through the existing old city along the Tigris River and the local authority allowed the construction of new three-storey buildings along its length with pedestrian colonnades or street arcades. In 1936, another modern street (Al-Kifah Street) was opened parallel to Al-Rasheed Street and along the site of the northern wall of Baghdad which had been demolished by Medhat Pasha in the nineteenth century. In 1956, a third wide avenue (Al-Jumhuriya Street) was opened between Al-Rasheed Street and Al-Kifah Street. Unlike Al-Rasheed Street, the local administration allowed higher buildings along Al-Jumhuriya Street with quite different architectural styles. Several crossing streets and roundabouts connected the three streets. These roundabouts were of various shapes and sizes and became landmarks in the modern fabric of the city, for example Al-Tahreer Square and Al-Rusafi Square. Most of the old buildings along these streets were demolished and replaced by new ones built in European styles, while the buildings and the streets located behind these modern streets were left untouched. The direct impact of these street projects was to divide the old city structure into different fragments, and because the streets belong to different decades, each has a unique architectural character.

From the 1950s the government initiated modifications along Al-Rasheed Street such as removing some structures to create new public squares and cutting or removing the balconies from buildings abutting the street so as to allow enough space for the London buses imported from England. Many buildings were also demolished and replaced with large and high-rise structures, so changing the established character of the street.

There are no reliable official data on Baghdad's population before 1965. However, at the end of the First World War, Baghdad was still a small capital city in a developing country, and had some 200,000 inhabitants. During the 1950s, the city began to experience pressure on its urban fabric similar to that in many other cities in developing countries. In 1965, the population reached 1.62 million (Hiwaish, 1988) and today it is estimated to be 5 million. This rapid population growth had major impacts on the city's infrastructure and created acute housing shortages. An early cause of these problems was an increase in the number of low-income rural migrants and their squatter settlements in the city. Yet until the 1950s, there was little planning to cope with the uncontrolled growth of the urban population and increasing numbers of squatter settlements. The many open spaces in the city together with the lack of central control made the squatting easy, and squatter settlements were not limited to Baghdad's fringes but also mushroomed in the open areas within the city. In the 1950s, Baghdad had some 44,000 *sarifa* (reed and mud houses) which accounted for about 45 per cent of the total number of houses in the city.[5]

Early efforts to develop a Master Plan were in general incomplete and fragmented as a consequence of the poor economy and lack of local town planners and architects. However, in 1927 oil was discovered in Kirkuk and increasing oil revenues contributed to improving the governments' ability to lay down new policies and projects for urban development. Among these were several housing projects carried out by English architects in Baghdad during the 1930s, such as Al-Sikak, Al-Rasmiyyah and Al-Elwiyyah.

In 1955, Doxiadis Associates were appointed by the Iraqi government to develop a plan for Baghdad. This was the first attempt to produce a comprehensive modern town plan for Baghdad to deal with its housing problems and urban growth.

Doxiadis's planning proposal for the city was based on implementing his principles of *ekistics* (from the ancient Greek meaning 'creating a settlement'). The method depended on an inter-disciplinary approach to the study of human settlements and the development of planning concepts and solutions based on this. His major argument was that human settlements were no longer satisfactory for their inhabitants (Doxiadis, 1968), and the problem was the imbalance between the different components of contemporary urban areas, which contributed to creating crowded and noisy cities which damaged both the natural environment and social life.

In his proposals for Baghdad, Doxiadis sought to solve the housing problem according to the inhabitants' economic means, their social life, the environment, and the local geography. His major concern was to manage the city's urban growth and to control its future development. The plan depended on leaving open spaces within the city, securing the expansion of the city core along a predetermined axis, so channelling urban expansion. The city was to be constructed in sectors, creating a modular grid. The make-up of the sectors was similar to a beehive, each with a central core with flexible basic building types (figure 4.2).

Between 1961 and 1963 many of the squatter settlements were eliminated and their inhabit-

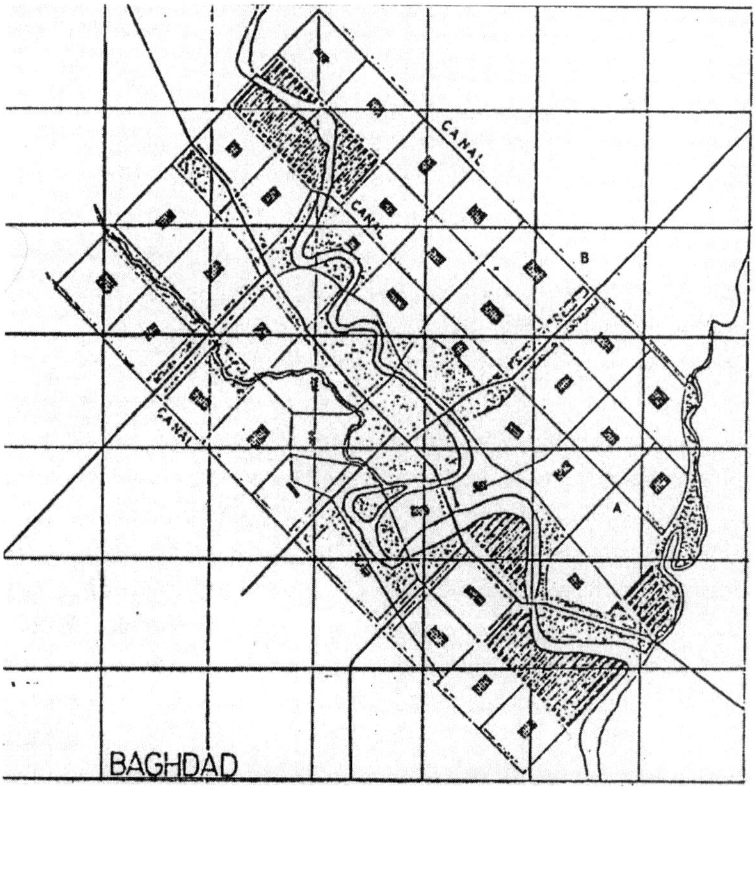

Figure 4.2. Town planning concept of Baghdad by Doxiadis.

ants moved to two large housing projects on the edge of the city at al-Nur and Madinat al-Thawra. Both were a direct result of Doxiadis's planning proposals. The planning and design principles of these neighbourhoods was based *inter alia* on the Iraqi family size and economy, but the plan proposals largely reflected Doxiadis's planning theory, which he had applied in many other countries.

In 1963, Iraq witnessed a military coup, which resulted in changes in local administration as people who were supporters of the new central political power were brought in. It is common, following a power change, for the new administration to change existing plans to prove that it is better than the previous one and to demonstrate its creativity and interaction with the new political system. The new local administration appointed the Iraq architect, Rifat Chadirji, to work together with Doxiadis Associates in order to develop their plans and design principles based on a better understanding of the local Iraqi needs. Shortly after that, the government decided to stop working with Doxiadis and abandon his planning proposals (Chadirji, 1986).

In 1965, the government asked new planning consultants, the Polish firm of Polservice, to develop alternative urban studies and plans for Baghdad to replace the earlier ones which were considered by the local authorities of Baghdad to be incomplete and unrealistic. According to the new agreement, Polservice developed a new master and zoning plan for Baghdad. The plan identified one main central area of the city along

Figure 4.3. The inner-city structure of Baghdad in 2000.

the two sides of the river Tigris. It also suggested developing eight sub-centres around the main core to decentralize the activities at the centre and so reduce pressure there (figure 4.3).

Among the major consequences of changing town planning principles from Doxiadis to Polservice were:

- the proposal for high-rise housing as a solution to the housing problem in Baghdad. It suggested increasing the number of high-rise buildings so as to house some 20 per cent of the city's population.
- the development of a new plan for public roads, classifying them according to new standards, such highways and main roads to solve traffic problems in the city and improve the urban development potential.

The government's strategic plan for Baghdad between 1975 and 1979 suggested prioritizing high-rise housing projects by increasing the number of residential flats by 70 per cent. These suggestions were adopted and realized in many large housing projects.

During the 1970s and 1980s Baghdad witnessed a boom in building construction in different sectors of the city, and on a scale not matched elsewhere in the country. The city became Iraq's architectural laboratory with different architectural styles being employed. Among the important projects of this period are the Al-Haifa Street and Abu Nawas development projects which reflect mixing international style and local architecture.

This period witnessed the increasing ability of local architects and other involved parties to participate actively in developing design projects through workshops, research, coordination, and conferences. Although, the process was still controlled by the central authority and influenced by its policies and priorities, these projects provide us with important lessons of how to create an interactive dialogue between local architects and international architects and others in order to develop modern design projects that match local needs. I will return to the architectural character of these projects below.

Changing the Traditional Architecture of Baghdad

Despite the general deterioration in Iraq during the late Ottoman period, traditional architecture managed to continue into the twentieth century (figure 4.4). This can still be seen today in some old neighbourhoods of Baghdad. These traditional buildings played an important role as a source of local architectural knowledge and design concepts.

Until the end of the nineteenth century, it was common for local craftsmen and those occupying and using buildings to function as designers. The local architecture of Baghdad was often based on vernacular architecture inherited from one generation to the next. The British colonial administration was the first to introduce architectural practice and Western style architecture to the city. This can be considered the starting point for the shift from local to modern architecture.

Colonialism established the basis for a form of open market in Iraq, which was helped by factors such as a population and economy large enough to support commercial enterprises in the country. In other words this was an early form of globalization, which had considerable impact on architectural practice and the type of architectural projects.

Many British architects established themselves in Baghdad, carrying out design projects which applied British architecture and imperial style mixed with some local design concepts. It was during this period that architects gradually replaced traditional Iraqi craftsmen. British architects also contributed to the importation

Figure 4.4. Old traditional houses in Baghdad.

of British and European building materials and technology in order to realize the new building designs. They carried out their projects without any local influence or control.

Most British and European design projects were for the government or the upper social classes (Chadirji, 1986, pp. 16–17). Among the buildings of this period is J.M. Wilson's Al-Albait University of 1922 (figure 4.5). It was also common for wealthy families to buy European house designs from local real estate agents who brought them from Europe, and then use local craftsmen for the construction (Chadirji, 1991, p. 517).

Iraq had no modern school of architecture to develop or absorb modern architectural trends and educate local architects. However, the increasing need for housing, services, and new types of building, particularly after the opening of the new streets and the rapid urban growth, increased the demand for architects. This demand encouraged the emergence, in the 1930s and 1940s, of the first generation of young Iraqi architects who were educated in the West, particularly in England.

Despite their small number, they slowly managed to influence local architecture in Baghdad by their projects and by participating in teaching new local architects. They played an important role in making the local community accept modern building technology and modern architectural movements. Fortunately, some of these early Iraqi architects were creative practitioners with a critical view of their own projects and of international architectural movements.

They used modern architectural knowledge and new building technology in their designs in order to meet the new needs of society. This was a natural consequence of their Western education and the absence of any local educational institution able to teach modern local architecture. The existing local architecture depended on old technology and its design principles suffered from repetition so becoming boring as a consequence of the stagnation in its development. As a result, until the 1950s there was no effort to study local architecture, to explore its conceptual character, or to search for methods to develop it or use it in new projects.

Forces other than architecture itself also influenced the shift to modern Western architecture.

British colonialism brought about direct contact with British culture. Because Britain was a global dominating power it was seen by Iraqis as a model from which to learn, particularly by young Iraqi architecture students who visited and lived there and admired its modern architecture and urban life. However, the imbalance between Iraqi and British culture created a gap between the local architects' desire to employ their imported design knowledge and the local reality of doing so.

In the absence of a modern local architecture and theoretical basis for it, the new generation of Iraqi architects had to depend on their own judgement as to how they should be influenced by Western architecture and by local traditional architecture. Their projects were experiments to determine the principles of a modern local architecture as yet unknown. The character of the new architecture was in principle pluralistic and largely depended on the ability of each particular architect to absorb modern architectural knowledge and reflect it in local architectural design and materials.

One can identify two different stages in the way in which local architects interacted with modern architecture in their projects. The first stage was the early projects, which depended

Figure 4.5. J.M. Wilson's Al-Albait University of 1922.

essentially on copying and repeating elements and concepts from imported Western architecture. Most of these projects reflected the international architecture of the time and many revealed a superficial understanding of modern architecture. The second stage depended on developing local knowledge so as to absorb, reflect, and create a new local architecture. In this stage we find local architects increasing their awareness and ability to learn from previous experiences, and using their own architectural heritage as a learning resource. From the 1960s this stage is also characterized by increasing debates and reflection on various architectural projects.

These debates and projects reveal the different methods Iraqi architects used to identify local architecture and use its design principles to create a modern Iraqi architecture. The central issues which concerned them were related to tradition, building materials, building technology, social life, and climate. Each architect emphasized one or more of these issues in order to identify the unknown local modern character. For example, from the 1970s tradition became an important issue as major urban development, road construction, and increasing land speculation in Baghdad resulted in the destruction of an important proportion of the historic fabric of the city. The local concern was how to improve these projects so as to ensure more harmony with Baghdad's local architectural tradition.

But the contemporary architecture of Baghdad is also a good example of how political change can bring about architectural change. The architecture of Baghdad is evidence of a phenomenon where form follows the political power system. This phenomenon has existed in Baghdad since the city was founded by the Abbasids in the eighth century.[6] During the twentieth century, Iraq witnessed radical changes in its political power system. The military revolution of 14 July 1958 replaced the monarchy with a republic. This power change was influenced by different local, regional and international forces, among them corruption, the creation of the State of Israel, the national revolution in Egypt in 1952, and the cold war. The revolution of 1958 was followed by two military coups in 1963 and 1968 that brought the Ba'ath party to power with its nationalist and socialist ideology. One of the symptoms of this political change in the Iraqi architectural movement was the increasing feeling of nationalism among local architects. The Architecture School of Baghdad was established in 1959 only a few months after the revolution.[7]

The Role of Local Architects, International Architects and Users in Shaping Baghdad's Contemporary Architecture

Different parties – local Iraqi architects, international architects, local builders and Baghdad's residents – have influenced the architecture of the city. Each had a different role in the way the urban fabric and architecture was created.

The Local Iraqi Architects

Until the 1990s local Iraqi architects could be divided into two groups: a small group who worked in the private sector and a larger group who worked for the government. Because of the government restrictions based on its socialist system, almost all architects educated in Iraq were obliged to work for the government.

Further, the Iraqi Architectural Association demanded 10 years experience before an Iraqi architect was permitted to establish a practice in the private sector. In the 1980s, the salary of a newly qualified architect was only 120 Iraqi dinar (about 350 US dollars) per month, which was relatively low compared to the cost of living. These conditions created a black market where young architects worked for existing architectural firms without any credit for their designs except payment. This unfair system also had negative effects on the architects' creativity, free competition, and the development of local architectural knowledge.

However, one can divide Iraqi architects' designs into two basic types: the first group follows the design principles of international architecture movements with little or no change; the second group seeks local design principles while at the same time applying contemporary international ideas. And this second group can be divided further into several levels depending on their definition of local architecture and their relation to international architectural movements.

Mohammed Makiya and Rifat Chadirji represent two important examples of Iraqi architects who played a key role in defining new Iraqi architecture. Both belong to the early generation of Iraqi architects, but they introduced two different ways to practice Western architectural design and how to deal with local architectural tradition. Ironically, both left Iraq and chose to live and work in Europe and the United States as a consequence of the political environment in the country.[8]

Rifat Chadirji[9]

Chadirji's efforts to find a local modern architecture developed through several steps. His early projects in the 1950s reflected strongly Western art and international architecture (Chadirji, 1985). After few design projects, he became aware of the gap between the technological development of building in the West and the local situation of Iraq. His conclusion was Iraq should have its own architecture that was an honest and creative outcome of the interaction between local technology, social needs and heritage.

His search for new modern local design principles led him to study and rediscover the Islamic and cultural heritage of Iraq. As a result he developed a distinctive Iraqi architectural style that he identifies as regionalized international architecture (Chadirji, 1986). In his designs, Chadirji depends on abstracting the concepts and elements of traditional buildings and reconstructing them in new forms. Among his projects are Tobacco Monopolies Administration, 1965 and Central Post Office, 1975 (figure 4.6).

His goal was the internationalization of traditional architectural concepts with the liberating design principles of modern architectural movements and the use of updated building technology. In almost all his writings, Chadirji argues that updated building technology is a vital factor in creating an updated local architecture. He criticizes, among others, Hassan Fathy's projects

Figure 4.6. Baghdad Central Post Office.

in Egypt (Fathy, 1971, 1973), which were based on vernacular architecture and used mud as the local traditional building material. He sees these projects as unrealistic ideas with a limited role in solving the present challenges in the Middle East city. It is an example of how to solve design in a rural village rather than a city.

Chadirji's arguments could be questioned, as Fathy's design method and theory aimed to use local knowledge and resources to produce architecture that fits the local need. They were not a general design model to be copied in other places, rather other regions should develop their own local design solutions based on the creative reuse and development of local vernacular architecture and building technology.[10] His was an attempt to develop a local alternative to contemporary Western architecture, which had made the local built environment dependent on it and Western building technology. In other words, Fathy's theory might today be identified as anti-globalization.

Chadirji designed a large number of projects in Baghdad and coordinated several important projects carried out by international consultants in the city. He therefore played an important role in the city's urban design and architecture.[11]

Analysing his Baghdad projects may reveal room for criticism. Despite his efforts to create a local modern Iraqi architecture and to consider local architectural tradition, he was largely concerned with creating modern local elevations rather than plan solutions. Therefore, we find in almost all his projects that the elevations contain local architectural character while the interior spaces are largely international style. This might reflect a method which deals with the shell rather than the content. The use of this two-way approach to design may be criticized because the plan is an important and interrelated part of a building's form and function. The search for modern local architecture should treat all aspects of design equally.

Chadiriji argues that the only alternative to creating a developed architecture is to bring the cultural development of Iraq into harmony with the process of internationalization in order to overcome the cultural gap during the development process, while at the same time maintaining the traditional character of Iraqi architecture (Chadirji, 1986, pp. 40–43). Therefore, regionalism, vernacularism, or nationalism cannot be successful methods to solve the problems of Iraq's interaction with the West. As an alternative he suggests regional internationalism as a solution.

He criticized projects by some Iraqi architects for being narrow and naïve in the way they copied concepts from Western or traditional architecture because they misunderstood the concepts and reused them in the wrong way. As a consequence their design concepts lost the original meaning. This point can largely be supported, but Chadirji's projects have similar problems.

In his method of abstracting traditional concepts and dealing with them as an art of reconstruction and recreation Chadirji lost their meanings. The beauty and aesthetic of traditional Baghdad architecture are largely related to its conceptual meanings. These conceptual meanings have a dialectical relationship with indoor and outdoor functions, which are reflected in the city's traditional structure, landscape and urban life.

Creating modern local architecture may require a better understanding of many issues such as the meanings of new design solutions and their impacts on social life in the streets and public spaces.

People's interpretation of the meanings of their built environment can be seen in the architectural solutions of buildings and public spaces, and the way they are used. For example, along the commercial streets of old Baghdad different types of structure and setting were built or created over time such as mosques, *sabils* (fountains), tombs, shops and *khans* (in whose courtyards traders from outside the city unloaded and sold their merchandise and found accommodation on the upper floors). Each particular structure and setting reflected its meaning in the street. During their everyday life, people could experience different meanings along the street such as God, death, identity, and community.

Mohammed Makiya

Mohammed Makiya was born in Baghdad in 1914. He studied architecture in England, at the Liverpool University School of Architecture in 1941, and then at Kings College, Cambridge where in 1964 he was awarded his PhD for his thesis 'Influence of Climate on Architectural Development in the Mediterranean Region'.[12]

Although Makiya was educated in the West and was influenced by its modern architectural theories and design principles, it did not take him long to be challenged by his own architectural reality in Iraq. He became increasingly concerned with issues of the local built environment such as the architectural heritage, fine art, folklore, climate, local building materials and local craftsmen.

Makiya depended on reusing the traditional design elements in new buildings and is often referred to as an Islamic architect (Makiya, 1990, p. 13). Although this definition may reflect different meanings, he revealed a deep understanding of and influence by traditional Islamic architecture.

In 1960 he designed the Khulafa Mosque in the centre of Baghdad (figure 4.7). This building can be said to be the start of his engagement with the problems associated with designing contem-

Figure 4.7. Al-Khulafa Mosque.

porary buildings and inventing local traditional architecture. The original Khulafa Mosque was built during the ninth century by the Abbasids. In old historical documents, it is referred to as Jamia al-Khulafa or the Mosque of the Caliph. For example, the medieval Moslem traveller Ibn Batuta visited Baghdad in 1327 AD and described this mosque as a very grand building and its surrounding area as modern and full of activity. When Makiya was awarded the contract to design the Mosque, the only part remaining of the orignal great Mosque was its minaret which was called Al-Ghazali and was located in the Sabbabigh al-Ali area of old Baghdad.

Although the project site was smaller than that referred to in historical documents, Makiya was able to develop a design extracted from the design concepts of traditional architecture. His method was based on using brick in association with concrete to reflect local identity. He used brick, which is the most traditional Iraqi building material, for decoration and to reflect the building's structural masses. In some projects, he was unsuccessful in using concrete to shape and construct traditional patterns such as the wooden *mashrabiyya* (pierced wooden screens) as they lost their original function, becoming merely decorative forms to give the elevation a traditional character. Among his other major buildings is the Foreign Ministry of 1967 which has elements influenced by Western architecture, particularly Boston City Hall (Kultermann, 1982, p. 55).

Yet, Makiya tried to bring a continuity of established local meanings into the contemporary built environment of Baghdad. One of the important characteristics of traditional architecture in Iraq is its ceremonial spaces and aesthetic effects that celebrate everyday life of the people. It was this effect that Makiya tried to create in his projects without neglecting modern building technology or contemporary architectural movements.

Makiya did not use only traditional Iraqi architecture as his source of architectural concepts but also architecture from other Islamic countries. In this way he considered the traditional Islamic world as an open and continuous architectural landscape, which had been produced by similar cultural influences. This reflects one of the controversial issues in his projects.

Study of the process that contributed to the creation of traditional Islamic architecture shows considerable exchange of architectural concepts between the different Islamic nations, as in the case of Baghdad and Cairo. Yet, a deeper study of traditional architecture in the different Islamic nations reveals that they are not homogenous. Their variety is a natural consequence of local conditions and materials. Their architectural principles have accumulated over long periods and created a traditional architecture peculiar to each Islamic society. Therefore, bringing traditional concepts from other Islamic countries into contemporary Iraqi buildings results in the same pitfalls as importing modern international architecture.

The International Architects

As mentioned above, the British architects who had settled in Baghdad following the First World War established the early tradition of using international architects in the city.

Improving economic conditions in Iraq encouraged local governments to develop the urban structure of Iraqi cities, particularly Baghdad. This development, which increased gradually from the 1950s and reached its peak during the 1970s, influenced both architectural education and practice. In the 1950s, the Iraqi government invited several international architects to design projects in Iraq. Baghdad hosted competitions and projects designed by architects from firms around the world. Among these were Le Corbusier, Walter Gropius, Frank Lloyd Wright, Pier Luigi Nervi, and Alvar Alato, each of whom belongs to a particular architectural movement. And in this way Iraq opened its doors to international architectural movements and not just British architecture.

Their contribution raised different controversial issues in relation to the city's holistic architecture, particularly the aesthetic relationship between the architectural character of the different designs and their relation to the existing traditional urban structure of Baghdad.

In a talk given by Frank Lloyd Wright at the San Rafael High School during July 1957, he said

> Now, at present, I happen to be doing a cultural center for the place where civilization was invented – that is Iraq. Before Iraq was destroyed it was a beautiful circular city built by Harun al Rashid but the Mongols came from the north and practically destroyed it. Now what is left of the city has struck oil and they have immense sums of money. They can bring back the city of Harun al Rashid today. They are not likely to do it because a lot of Western architects are in there already building skyscrapers all over the place and they are going to meet the destruction that is barging in on all big Western cities. So it seems to me vital over there to try and make them see how foolish it is to join that Western procession.

Indeed, Frank Lloyd Wright expressed a sensitive concern about international architects' involvement in changing the urban fabric of Baghdad. Two decades later the dimension of these changes in Baghdad's architecture was all too evident. Frank Lloyd Wright's design proposal for the University of Baghdad was criticized and neglected by the local authority. Another location was chosen and in 1961 the design was given to Walter Gropius and his TAC group.

It was not until the 1970s that most of the gigantic projects were designed or realized. Inviting famous architects to design several spectacular architectural projects was to increase the rate of development in order to put the architecture of Baghdad into line with that of developed countries' capital cities. This policy was called *jumping policy* (*siyasat altafra*) and included every sector of society. It was encouraged by the economic development as a result of increasing oil revenues following nationalization of Iraqi oil and the sharp rise in oil prices.

During this period, the nature of architectural discussion had broadened to involve larger issues such as urban growth, social community life, environment, conservation and heritage. Baghdad's new architectural projects became an interdisciplinary field where experts from different disciplines began to examine the city's contemporary and future urban design and architecture. But all were constrained by the limits of the existing city fabric and its planning which had been developed in the early twentieth century.

The jumping policy and its architectural projects raised strong debate among local architects. The required speed of urban change was faster than any expectation or imagination and it lacked the critical view of how society could absorb such rapid change. However, most of these large planned projects were never realized, particularly after 1984, because of the increasing costs of the war with Iran and the political instability that left the country with serious economic problems.

Some of the projects of this period are: the Saddam Hussain Gymnasium designed by Le Corbusier in 1964 and built in 1980; the Central Bank of Iraq, 1985 by Dissing and Weitling; Baghdad Conference Palace, 1982 by Heikki Siren; Baghdad University City designed by Walter Gropius and his TAC team; Haifa Street Public Housing, 1984 designed by ARC Design Consultants Ltd; and the Abu Nawas Development Project, 1981 by Skaarup & Jespersen (figure 4.8).

Although Iraq was exposed to a strong international architectural movement during the 1970s and 1980s, this was a process controlled by the central authority of the Ba'ath party government. The different authorities such as ministries and the municipality (*amanita alasimah*) identified the development projects and their dimension, selected project committees and controlled the entire process. The process was often governed by the ideological and political motives of the ruling Ba'ath party. A further problem was the absence of a group of local architects with homogenous understanding of local architecture. This can be seen in the architectural design projects of Chadirji and Makiya mentioned above. Therefore, the local Iraqi committees and architectural firms

Figure 4.8. Abu Nawas Development Project.

who joined the international architectural firms as advisors or partners led them in different directions. The absence of a common platform of local architectural principles is reflected in the range of architectural styles exhibited in the design projects.

It was also during the 1970s that ordinary Iraqis' incomes improved thus encouraging construction activities in the private sector, such as houses, flats, hotels and commercial buildings. It became normal to turn the houses or the front gardens of large houses along the main streets into commercial buildings. These projects were often carried out by local architects and reflect a different level of architectural design quality and character.

The People's Role

Ordinary people had an important role in shaping Baghdad and its architecture by their individual actions or together with the cooperative unions. This was through their participation in construction activities, in applying individual design concepts, and in modifying the existing buildings and open spaces. This involved houses and small commercial buildings, which were often designed by the inhabitants and local builders. They might copy design solutions from existing buildings or plans done by others. This phenomenon was inherited from old customs, which existed before the emergence of modern architectural practice. But the traditional built environment provided its inhabitants with design solutions that were developed over a long period and their meanings were simply understood. Therefore, these design principles were deeply rooted in local knowledge, which enabled ordinary people to reuse and change their old design principles to meet new needs more easily and in a more creative way than their present efforts of copying from contemporary architecture.

The role of the people is also evident in changing the design and use of spaces and their architectural character. This trend can be seen, for example, in the occupation of sidewalk areas by the shop owners and residents who use these spaces for new functions such as displaying goods, sitting, playing, and building front gardens. This transformation is also evident in changes in the function and/or design solutions of the existing buildings and the use of the front gardens of large houses along main streets to build shops and commercial buildings. This process of creating new settings in the existing built environment is continuous, frequent and rapid. When a transformation in one place is seen to be beneficial, it can quickly be interpreted and transformed to another within the city.

People's active participation in modifying and creating new architecture and settings is either a means to solve some problems of the designed buildings and open spaces or in response to other forces such as population growth, increasing commercial competition, climate, increasing land prices, or changing lifestyle. The participation of the inhabitants in changing the built environment of Baghdad involves all social classes. Their response to the planned city and its architecture has modified the built environment in such a way that they can practice their traditional way of life and reflect their lifestyles in the built environment.

For several decades following the 1940s the people also played a major role in cooperative housing projects in Baghdad. The purpose of this housing was to provide accommodation for the working classes and improve their living environment by joint action between the state, the cooperative, and individuals. The projects contributed to the establishment of communities and neighbourhoods, where each neighbourhood housed a particular group of workers, such as engineers, doctors, officers, and so forth.

The Architecture of Baghdad under International Sanctions

We know today, as a consequence of accumulated experience, that Baghdad's architectural beauty is not only due to its well developed local architecture, but also to the city's historical charm, which is deeply rooted in the rich history of the Islamic period.

It is often referred to in Western literature as 'the city of a thousand and one nights' and might reflect the hidden story of an old city that was full of life and historical events. It has fallen and been raised several times in its history. At the time of writing, it is probably the most media covered city in the world, following the political conflicts between Iraq and the United Nations, three Gulf Wars and continuing unrest.

Since 1990 Iraq has been under tough and controversial international sanctions by the UN as a result of its invasion of Kuwait in August 1990. The Gulf War in 1991 caused large-scale destruction of basic infrastructure and buildings. Even historic buildings, sites, and ruins suffered rocket attacks by the international coalition forces. During the forty-three days of that war, the city was bombarded with weapons ranging from laser-guided bombs to Tomahawk cruise missiles. People's life support systems – electricity, water, transportation, and communications – were disabled. Among the targets were the power station, government conference centre, TV transmitter, communication relay station, Jumhuriya Bridge, Karada Bridge (14th July Bridge), telephone switching station, and the Al-Dawrah oil refinery.

The earliest efforts by the local authority were to rebuild the infrastructure and the basic government buildings. The most tragic comedy of this period was the rebuilding of the presidential palaces and even building new ones while Iraq suffered under UN sanctions. At a time when all available money was needed for reconstruction, the government used millions of dollars to finance these palaces. The presidential buildings provide us with a good example of how Iraqi architecture is responding to the sanctions, to globalization, and to the associated political conflict in Iraq.

The Iraqi architectural phenomenon of this period would best be called 'architecture in the era of sanctions'. The vast building constructions, which were common during the 1980s have disappeared. During this period, one can note several directions in the local architecture of Baghdad. The first trend is reflected in the small projects carried out by individuals. Here one sees imported European design elements used in the design of houses and small buildings, which often include features from ancient Greek and Roman architecture, particularly columns.

The reason for this phenomenon can be attributed to several issues, among them the socio-economic conditions under UN sanctions and their associated psychological effects. The sanctions caused a dramatic change in the nature of existing social classes whereby the middle class has almost disappeared and a new rich class emerged. During this period it became normal for architects to import elements from neighbouring countries, particularly Jordan, Syria and Turkey. The Internet has also played an important role, enabling local architects to reflect or even copy architectural design elements in new buildings. There is little critique of this and it may reflect the general conditions in a society where most people are more concerned to secure their economic life and basic everyday needs under the sanctions and now under the present occupation. As a consequence of the economic and political situation in the country, many Iraqi architects have left Iraq since the 1991 Gulf War and settled in other countries. Their absence reduced local architectural debate and creative interaction with imported architectural styles. In many ways the present architectural situation in Baghdad is similar to that during the early 1940s. On the other hand, UN sanctions against Iraq have made the Internet an important source for updating the architectural knowledge of local architects and students.

The second trend in local architecture was encouraged by the ruling power and depends on using architecture as a tool in the political challenge to international powers and as local propaganda to demonstrate the power of the ruling system. For the government to build anything at all was a challenge, but they chose large buildings with rich traditional design elements as evidenced by the presidential palaces and state mosques. Among such projects is the Um al Maarik Mosque (Mother of All Battles Mosque).

In a talk given by former president Saddam Hussein to Iraqi architects in June 2001, he advised and asked them to purify their designs of Western architecture and to concentrate on truly traditional Iraqi architecture. His justification for this was because original Iraqi architecture is connected with the country's civilization.[13]

This architecture, created as a demonstration of power, has been practiced since the 1980s following the Iraq-Iran War. Among the projects are memorials to the Iraq-Iran War such as the Tomb of the Unknown Soldier, the Martyr's Memorial, and the Triumphal Arches. These projects were designed as vast structures and occupied large open spaces.

Constructing presidential palaces under UN sanctions, thus using architecture as an argument in international conflict, created one of the most unique architectural phenomena in the history of Iraq or perhaps the whole world. The size and architectural quality of these palaces increased their construction costs and caused much criticism from the US government against the Iraqi government. The United States claimed that Saddam had spent a billion dollars between 1991 and 1996 on palace building instead of using the money to meet civilian needs. Although the figure of one billion may be exaggerated, the construction of these lavish palaces took place in a period when ordinary Iraqis were suffering shortages and crisis.

These palaces were suspected of housing biological and chemical weapons and caused several political conflicts between Iraq and the United States because the Iraqi president blocked United Nations teams from inspecting them. Military conflict was averted temporarily by the Iraqi government's decision to permit the inspection of the presidential palaces and other sites in Iraq by UN inspectors following UN resolution 1441 of 8th November 2002.

Rebuilding Iraq and the Architecture of the City

On the 20th March, 2003 the coalition forces of the United States and United Kingdom started a major new war to remove the Iraqi regime and remove weapons of mass destruction from the country. They attacked large numbers of sites including Baghdad and other major cities using all kinds of modern weapons, targeting and destroying major buildings and other constructions. On the 9th April 2003, Baghdad finally fell to the coalition forces and the city witnessed further damage from the looting and burning of large numbers of public and governmental buildings (figure 4.9). Such looting included stealing most of the Iraqi traditional treasure from the Iraqi Museum and the Iraqi Archive Centre. Although many of these treasures have since been returned, many others are still missing.

The American and British governments have repeatedly announced that their efforts will be concentrated on rebuilding Iraq and creating a new democratic environment.

At present, Iraq is controlled and administrated by the coalition forces led by a US civil administration. Recently, a transitional political system was established to participate in the reconstruction process and the creation the new Iraq. It is still early to talk about the outcome of this process and its effects on the architecture of Baghdad and other Iraqi cities, but building the democratic process in Iraq and applying it to construction activities and the architectural debate are vital issues in this phase. There are important lessons that the US civil administration and the transitional Iraqi government should learn in order to establish the right local environment for the rebuilding process. These come from the accumulated experiences of Iraq's modern history, and the shortest way of learning is from the lessons

Figure 4.9. Baghdad Central Post Office in April 2003.

of Baghdad's architectural development during the twentieth century.

Conclusion

The historic development of architecture and town planning in Baghdad clearly embodies different ways of thinking about the city's planning and its architecture. It also embodies different ways of influencing and responding to the rules, logic, and pressures of the changing local and international market system and the changing ruling powers during the twentieth century.

From the 1920s, the Iraqi construction market increasingly revealed a strong client-based pull factor, which attracted large numbers of international architectural and construction firms. The contemporary character of Baghdad's architecture was developed over time by local and cosmopolitan efforts of architects from various nations and architectural movements. The outcome of this process is a city with a variety of architectural characters. Although each particular building or project may have good architectural design principles, the overall result is one of diffused architectural and urban space.

The architecture of Baghdad during the twentieth century was always influenced by the ruling power system and the economic situation in Iraq. This influence varied from weak to strong intervention by the central authority in the process of choosing construction projects, and the architectural firms, both local and international, to carry them out. The political instability and several power shifts during the twentieth century also contributed to shifting Baghdad's town planning principles and the concepts of the city's Master Plan. The urban development strategy of each particular government had different programmes and priorities, which contributed further to diffuse Baghdad's contemporary architectural character. I use the term 'diffused' rather than 'diversified' because almost every area in Baghdad developed its own character and the only relationship between areas is the street network.

The central authority's policy restricted newly qualified architects from participating in the private sector; they could only work for the government. This restriction had direct consequences on the quality of architectural production and the development of local architectural knowledge.

The Ottoman legal system, which originally regulated town planning, construction activities and building, was replaced by a Western system, based mainly on British and French law. This shift took place without consideration of the consequences for the use and design character of the urban fabric. No attempt was made to develop local alternatives for these regulations. It is hard to create local urban design concepts and local architecture without developing building codes and regulations which reflect local needs and cultural heritage.

Globalization, too, can have impacts on the process of creating and developing local architecture. Domination of the architectural market by international architectural firms without local control can create many problems for society. Local architecture should not be understood merely as an industrial product. To combat this requires the development of flexible measures to counter the effects of globalization on local architecture and urban design. These should contain components to organize international firms' involvement in the local architecture market through, for example, the development of local building codes and local architectural knowledge, and the involvement of local architects in the process.

Such measures should establish a deeper

understanding of local architecture's roots, its growth and its relationships with the social life, resources, and local requirements. Local architecture is developed by the accumulation of local knowledge from everyday life experiences. Observing, studying, and understanding this should be the basic issue for any type of design.

Therefore, in an age of globalization, the dominance of international firms in local architectural markets should be avoided. As an alternative, the national system should support and contribute to building local architecture and local architectural firms in parallel with establishing cooperation between local architects and international firms to stimulate better solutions and build up local and international architectural knowledge. The important issue should not be local architects or international architects but the local architecture and how to develop it. Some 1980s projects in Baghdad provide an interesting lesson to support local architecture in the era of globalization.

A general observation of Baghdad's architecture shows that Iraqi architects developed different methods of creating local architectural identity while reflecting modern architectural movements in their designs. These fall into three basic types:

- using concepts and principles from international architectural movements with almost no change;
- using modern design solutions combined with some abstracted elements and concepts from local traditional architecture;
- using design concepts and elements from traditional architecture as a basic source of contemporary design solutions.

The search for local architecture had paid little attention to developing local urban design. The town planning models that were applied could have been applied in any city or country. There was more concentration on building design than on identifying urban space design elements, which constitute Baghdad's urban character. For example, there was little attempt to create street designs with traditional character which were sympathetic to local people's social activities on the street.

This may also require a new way of understanding and using tradition in modern local architecture. An alternative to the present way is making use of tradition in architecture that is relevant and not merely nostalgia or the recreation of old tales. Using tradition should not be limited only to abstract and single building design elements but should also involve other components such as space design concepts, use and meanings. Understanding the relationship between these issues and their role in the culture of local society means one should include wider cultural issues in architecture. The challenge is how to study and use traditional aspects by exploring their reality without distortion or misrepresentation. The challenge, too, is to suggest innovative ways for their application not only for elevations but also indoor and outdoor spaces, streets, and in urban design as a holistic composition.

The holistic approach suggests better understanding of the users' role in creating local Iraqi architecture and urban design. The newly created settings and modification of existing architecture have contributed to the diffused character of contemporary architecture in Baghdad. This can be used as a source for developing knowledge about the local built environment, urban design, architecture, policies, and their related rules. It requires identification and an understanding of the forces at work in the local urban environment – culture, central control, population growth, and socio-economic class, privacy, economic competition, climate, pollution, and traffic – and how people use their tradition to interpret the urban spaces in the context of these forces. The designer should reflect on these issues in future design and town planning.

We have seen how, despite the formal zoning of Baghdad and construction of modern buildings and urban spaces, people intervene to modify, adjust and create new or neglected functions and settings in order to balance and regenerate their living environment. This is a general phenomenon in both the commercial and residential areas of all social classes in Baghdad and is influenced largely by the people's culture.

This phenomenon shows that architectural reality does not necessarily end by shaping the building. Shaping the building is only the starting point of a series of reshaping processes of the building's form, function, and its surrounding spaces. The nature, dimension and speed of the reshaping process is largely influenced by local forces. Therefore, buildings and urban spaces should be created in familiar dimensions and character based on lessons from local tradition.

It is the democratization of the architectural process that is needed in order to build local systems capable of dealing with new issues related to globalization so as to form well thought out architecture and far sighted strategies for the transformation of Baghdad and other cities in post-war Iraq.

NOTES

1. The Islamic Calendar started in 622 AD when the Moslems settled in Medina and is referred to here as 1 H/622 AD.

2. See for example: http://www.btinternet.com/~britishempire/maproom/iraq.htm.

3. In some studies Imam Shafi'i is presented as the founder of Islamic law (*shari'ah*). This can simply be refuted because Imam Abu Hanifah died in the year H 150/767 AD which was the same year Imam Shafi'i was born. See also Imam Abu Zahrah, Mohammed (no date).

4. *Tanzim* plural *Tanzimat* (reorganaization) were plans introduced in the Ottoman Empire to reform the administration and legal system during the nineteenth century. See also Lewis (2001).

5. See, for example, American University (1979), chapter 2.

6. See also AlSayyad (1992).

7. The Department of Architectural Engineering at the University of Baghdad states in its curriculum of 1983–1984 that 'Through the use of distinctive philosophy and courses program, the Department aims to establish and crystallize a local architectural trend that has its own cultural and technical character. The local trend created by the combination of the rich architectural heritage of the region and current technical and scientific development'.

8. Hundreds of Iraqi architects have fled Iraq since the 1980s and settled in different countries around the world. Rebuilding of Iraq should include plans to encourage those experienced architects to return to their homeland and to involve them as active participants in the reconstruction process. See, for example, http://www.iraqiarchitects.com.

9. Born 1926 in Baghdad, educated in architecture in England, he practiced architecture in Baghdad from 1952. He designed a large number of well-known projects in Iraq and the Middle East. He also has many publications in Arabic and English.

10. Although it was called architecture for the poor, its design principles were largely used in tourist sites, palaces, and other expensive buildings.

11. See, for example, the studies made by TAC (The Architects Collaborative) in Cambridge (USA) of the Khulafa Street projects in Baghdad in Bianca (2000), pp. 253–269.

12. Kufa Gallery (2002) (http://www.kufa.dircon.co.uk/Dr_Makiya/Biography.htm).

13. http://www.uruklink.net/iraq/e2001/espeech1_6.htm.

REFERENCES

Alsayyad, Nezar (ed.) (1992) *Forms of Dominance: on the architecture and urbanism of the colonial enterprise.* Aldershot: Avebury.

American University (1979) *Iraq: A country study*. Washington D.C.: The American University.

Barzinji, Saedi (1978) Clause d'exonération de la responsabilité contractuelle en droit Francais et en droit Irakien (Discharge conditions from contracting liability in French and Iraqi Laws). Published PhD Thesis, Université des Sciences Sociales, Grenoble-France.

Barzinji, Saedi (1989) Arbitration on the Conflicts Emerging from Civil Engineering Contracts. Paper presented to the 1st Scientific Conference of the Work and Housing in Iraq. Baghdad.

Bianca, Stefano (2000) *Urban Form in the Arab World*. London: Thames and Hudson.

Chadirji, Rifat (1985) *Taha Street and Hammersmith*. Beirut: Muassassat Alibhath Alarabiyah.

Chadirji, Rifat (1986) *Concepts and Influences: Towards a Regionalized International Architecture*. London: KPI, pp. 16–17.

Chadirji, Rifat (1991) *Al-Ukhaider and the Crystal Palace*. London: Ryad El-Rayyes Books Ltd, pp. 88–92.

Doxiadis, Constantinos (1968) *Ekistics: an introduction to the science of human settlements*. Oxford: Oxford University Press.

Engardio, Pete *et al.* (2003) The new global shift. *Businessweek*, February 3, pp. 1–7.

Fathy, H. (1971) The nature of Islamic urbanization: a historical perspective. *Ekistics*, 182, pp. 62–63.

Fathy, H. (1973) *Architecture for the Poor*. Chicago: University of Chicago Press.

Hakim, Besim (1971) The nature of Islamic urbanization: a historical perspective, *Ekistics*, No. 182, pp. 62–63.

Hakim, Besim (1972) Coop housing, Baghdad: an evaluation and recommendations. *Ekistics*, No. 196, pp. 166–172.

Hiwaish, Akeel (1988) *Modern Architecture in Iraq*. Baghdad: Dar al-Shoun al-Thaqafiya (in Arabic).

Imam Abu Zahrah, Mohammed (nd) T'arikh al-Madha'hib al-Islamiyyah (History of Islamic Jurisprudence) (in Arabic). Cairo: Dar al-Fikr al-'Arabi.

Kultermann, Udo (1982) The architects of Iraq. *Mimar Journal*, No. 5, p. 55.

Lasner, Jacob (1970) *The Topography of Baghdad in Early Middle Ages – Texts and Studies*. Detroit: Wayne State University Press.

Lewis, Bernard (2001)*The Emergence of Modern Turkey*. Oxford: Oxford University Press.

Makiya, Kanan (1990) *Post Islamic Classicism*. London: Saqi Books.

Morris, A.E.J. (1994) *History of Urban Form*. Harlow: Longman Scientific & Technical, Harlow.

Nooraddin, Hoshiar (1996) *Al-fina': A Study of 'In-between' Spaces Along Streets as an Urban Design Concept in Islamic Cities of the Middle East with a Case Study in Cairo*. Trondheim: NTNU.

Stiglitz, Joseph E. (2002) *Globalization and Its Discontents*. New York: Norton.

ACKNOWLEDGMENTS

I would like to extend my appreciation to my wife Tahany and our lovely children Suzan, Noor, Jean and Imran for their wonderful support and help in collecting important information during my work on this chapter. I would like thank Professor Besim Hakim in New Mexico and Professor Rifat Chadirji and Chadirji Research Centre in London for providing me with important advice and photos. I would like also to thank my brother Saedi and my nephew Zakarrya for assisting me during my visit to beautiful Baghdad.

Chapter 5

Sana'a: Transformation of the Old City and the Impacts of the Modern Era

Khaled A. Al-Sallal

Sana'a is the city of Yemen – there is not to be found in the [highland of] Yemen, nor in the Tihamah nor the Hijaz a city greater or more populous nor with finer properties, nor any nobler of origin . . .

Ibn Rustah, *Kitab al-Alaq al-nafisah*, written after AD 903

The old city of Sana'a possesses a uniqueness in Yemen and in the world which can only be compared to the special quality of Venice; that is, its value lies not so much in the merit of the individual buildings, important though they may be, as in the unforgettable impression made by the whole – an entire city of splendid buildings combining to create an urban effect of extraordinary fascination and beauty.

Ronald Lewcock, *The Old Walled City of Sana'a*, 1986

Prologue

From vertical to horizontal design; from communal spaces to individualized spaces; from the use of local materials and technologies to the anonymous – these are some of the extreme transformations that have taken place in the architecture of Sana'a during the last 40 years as a result of several important political, economic, and social forces. The vertical forms that gave Yemeni architecture its unique identity have been transformed into horizontal forms that have no roots or identity. The communal green spaces transformed into individualized, mostly concrete paved spaces. A sophisticated Sana'ani window design, known by its separated components that gave great flexibility in use, transformed into endless types of windows with no consideration of order or climate.

Once famed as the 'Pearl of Arabia', the old walled city of Sana'a in south-western Arabia has been continuously inhabited for over 2,500 years (figure 5.1). It is one of the world's most ancient surviving cities and, arguably, the world's longest continually inhabited city. It was nominated to be among new seven wonders of the world by Bernard Weber, a Swiss filmmaker, who has launched a web-based poll[1] to define seven new symbols of humanity's greatest achievements of the last 2000 years. 'A remarkably impressive architectural wonder and one of the most significant historical sites in the Middle East', Weber said of the old city of Sana'a.[2] The city is dominated by

Figure 5.1. Panoramic view of Sana'a.

6,500 tower houses of five to nine storeys, and its religious and cultural heritage is reflected in its 106 mosques and twelve *hammams* (public bath houses) built before the eleventh century. The houses were built of baked brick, and decorated with gypsum and dressed stones, in a marvellous and extraordinary architectural ensemble. 'The city's architecture has been damaged, demolished and rebuilt through flooding, wars and prosperity. Yet, it wasn't until the modernization in the 1970s that the city's architectural fabric was truly in danger of disappearing.'[3]

Yemen entered a totally new era in its history after the establishment of the Yemen Arab Republic in 1962. This new political regime focused on establishing the foundations of modern Yemen in aspects such as education, health, and economy, that had been totally neglected for centuries during the Imamic ruling period. With equal importance, it also opened doors that had been closed for a long time by establishing political and economic relationships with the rest of the world at regional and international levels. Other important historical events that have had a significant impact on the society and economy of Yemen are the unification of North and South in 1990, the Gulf War (of 1991), and the Civil War. The impact of these events on the rate of labour migration from Yemen to Saudi Arabia and other Gulf countries have had a significant economic impact on the urban and architectural patterns of Sana'a. Also, the migration from the Yemeni villages to the capital had a major effect on the urban and architectural patterns.

Since the Yemeni Revolution in 1962, remarkable transformations in urban and architectural forms have been taking place as a response to enormous changes in the politics, economy, and society of the country. These transformations were caused by a large number of factors with complex and multi-layered variables and relationships. Unfortunately, the new architecture that emerged after 40 years of transformations is mediocre except for few cases, compared to the vernacular architecture, and is not satisfactory for most people for the following reasons:

- It is weak and has no identity or belonging whatsoever; it is a cheap mix of Yemeni and Western styles and lacks cultural meaning. To the contrary of this, the vernacular architecture of Yemen is known for its strong character, remarkable identity, and deep cultural meaning.
- It lacks order and unity at all levels: from detailed level of form and façade design to the space design level, and from the selection and use of materials to the construction methods employed.
- It is alienated in its society because it provides minimum consideration to socio-cultural needs such as privacy and sociability.

- It is alienated in its environment because of its negative effect on the environment. It is not environmentally sustainable and its reliance on, or integration with, natural resources of energy, vegetation, or water is minimal.
- It gives little or no consideration to human comfort issues, especially thermal comfort.

To realize this transformational process, there is a need not only to reveal the values of the vernacular architecture, but also to understand how the vernacular architecture succeeded in keeping a balance among complex relationships of environmental and socio-cultural variables; and why the modern architecture failed to do so. Why has vernacular architecture in places like Sana'a proved to be more successful than contemporary? To answer this commonly asked question, we need to understand the intricate relationship between culture, environment, and architecture in the traditional society of Sana'a, a society that is well known for its rich cultural values.

Most of the relevant literature available today has been produced by scholars with non-architectural or design backgrounds and has focused on the history of Yemeni architecture, describing the physical built environment with little relation to the cultural and ecological context. Al-Sabahi (1996) argues that most of the literature was concerned with the history and architectural style of Sana'a, either in itself, or in the context of Yemeni or Arab-Islamic architecture. His study also argues that a deficiency of this literature has been a lack of detailed analysis of the architectural and social values inherent in the concept of the housing cluster as it is found in the old city, and the way that these values have been transformed with the construction of the new suburban areas since 1962. Also, with a few exceptions discussed below, most studies have neglected the history-long valuable bio-climatic experiences that can be learned from the Yemeni architecture and have not considered its important relationship with other social and cultural aspects.

A major study that describes the old walled city of Sana'a in detail is Serjeant and Lewcock (1983). This focuses on the historical background and development of the city, starting from the pre-Islamic time. It gives a historical description of important buildings and elements such as the major mosques, the *hammams*, the *souq* (the market), and a description of the housing clusters. It also describes the social and daily life of the Sana'a people, their habits, business life, traditional dress, and food. In spite of its comprehensive description of many important historical details and social patterns, it does not analyse the nature of the changes that have taken place with the development of the modern suburban areas of Sana'a. Another study by Lewcock (1986) with support from UNESCO explains the historical background and development of the city, the social structure of its people, and its architecture. It also explains the dangers threatening the future of the old walled city and the problems confronting its inhabitants. Among the problems identified are the poor water and drainage systems, the difficulties in maintaining old buildings, and the lack of public amenities. The main objective of the study was to produce a strategy for the conservation of the old city of Sana'a.

Marechaux (1987) describes several important points about Sana'a history during ancient and Islamic times. He highlights the holy status of the city among the Southern Arabia civilizations and its significant role as a fortified city. He describes the urban pattern of the old city of Sana'a and discusses the shortcomings of the old city's infrastructure, comparing it with the modern suburbs that are provided with amenities. He also describes the changes that emerged after the Civil War with regard to social and economic aspects.

After discussing the vernacular housing in Yemeni cities including the styles, materials, and building techniques,Veranda (1987) describes the modern housing styles that emerged in Yemen after the Civil War. As quoted by Al-Sabahi (1996), Veranda points out:

> With the end of Civil War, new house types were introduced: urban apartment buildings built in concrete up to four stories high and single family houses modeled on the villa, with one or two stories. The change in social and economic status identified with these houses is reflected in a somewhat disordered ornamental display, both in color and texture.

Hirschi and Hirschi (1983) introduce a comparative analysis of architectural styles in several locations in Yemen such as: Sana'a, Dhamar, Jiblah and Zabid. The study provides a comprehensive analysis of vernacular architecture in terms of local styles, pointing out factors such as climate variations, readily available building materials, ornamental patterns, and the arrangement of windows. As they point out, because of pressures of social and economic change on Yemeni society, the importance of taking urgent action to conserve the vernacular architecture should be taken seriously. They also believe that Yemeni architectural heritage should be preserved as a valuable resource for designing new building projects.

Matthews (1979) argues that the old city of Sana'a had satisfied the needs of the people until recent times. Today's problems such as traffic congestion were never imagined when the city was built; these problems are new and require new solutions. Once the isolation of Yemen was ended in 1962, town planners copied Western styles and used imported building methods and materials during a time that lacked planning legislation and building codes.

In 1985, UNESCO allocated funds through a long-term plan to save the old city of Sana'a (Rogers, 1986; Al-Sabahi, 1996). The objective of the plan was to arrest actions threatening the architecture of the city of Sana'a. Pollution and the difficulty of driving cars through the narrow streets of the old city had been causing serious damage to the buildings. Money was allocated to pave roads and build parking facilities so as to keep cars out of the old city. Also, traffic management was established. Some of these efforts were successful; yet because of the high costs of building materials and labour, some owners were not able to repair their damaged houses. Most of UNESCO funds allocated for renovation were used on public buildings. A major problem occurred when the UNESCO plan attempted to provide a new system for water and sewage with increased flows; the ancient cess-pits could not cope with this and cracks and leakages into the city's water table resulted. It also threatened the structural stability of many buildings.

Al-Sabahi's (1996) comprehensive study describes the architectural style of old Sana'a using three levels: the urban structure, the traditional housing cluster, and the vernacular house. He explains the urban administration and development of Sana'a during the late Imamic regime (1918–1962) and under the Republican regime, and discusses the impact of modernization and socio-economic change on Sana'a during two distinctive periods 1962–1970 and 1970–1991. He provides an interesting comparative analysis of the traditional and modern sectors in Sana'a with regard to land use and architectural expression using traditional and modern case studies.

A study by Al-Oulfi (2000) analyses Yemeni architecture in different locations by giving a comparison of architectural forms and elements. He also provides a historical background in which the Yemeni architecture is compared and related to other architectures in the Arab region and in the world. He describes the modern movements in Yemeni architecture and discusses the issue of how a sustainable architecture can be created in the country in the modern era.

Despite the richness of its environmental and socio-cultural design solutions, the architecture of the southern Arabian region has not been widely researched. A previous study shows that the vernacular house in Sana'a provides better thermal comfort than the modern house (Al-Sallal *et al.*, 1995); yet very few of these buildings have been built in the last two decades. That is because of the dominance of Western design and materials which were not developed to meet the bioclimatic and functional needs. Al-Sallal (1993, 1996*a*) investigates the role of indigenous fenestration to reduce energy requirements in the Sana'a house using computer simulation. Another study by Al-Sallal and Cook (1992) documents the indigenous window of Sana'a and analyses its design variations in size, shape, and number based on climatic influences and functional requirements. They categorize the window design according to its components and describe the geometry and function of each component. They conclude that the separation of window into components gave better flexibility in controlling the different functions and more design opportunities. Al-Sallal (1996*b*) employed three-dimensional computer modelling for visualizing solar access and shading in the traditional housing cluster of Sana'a historical city. He found that their urban pattern, orientation, and the houses' vertical forms helped to maximize solar access in the winter from the south, while avoiding it in the summer. In two different studies, Ayssa (1990, 1994) investigates the traditional window shutters and attachment as comfort producing procedures and the internal air movement in the vernacular houses of the old city of Sana'a. Ayssa (1996) describes the passively cooled vernacular Zabeedi house and the potential to improve it.

In a recent study, Al-Sallal (2001) investigates the balanced synthesis of form and space in the vernacular house of Sana'a, and how that was used as the chief means to mitigate the effect of the climate and provide the functional requirements of culture. The approach depends on a number of methods for analysis of common architectural practices such as passive solar thermal performance analysis, solar access and shading analysis using 3D computer modelling, and observation and literature survey of cultural aspects. A summary of the study's outcomes provides rules of thumb that can be applied in design.

Background

Geography and Climate of Yemen

Yemen, the very south-west corner of the Arabian Peninsula, is situated between latitude 12° 40′ and 17° 26′ north of the equator and longitude 42° 30′ and 46° 31′ east of Greenwich. At first glance it seems to be the continuation of the desert that exists in most of the Arabian Peninsula. Yet Yemen has a very different climate, mainly because of its altitude. Here one finds the highest mountains of Arabia, rising to 3,720 m. Large valleys above 2,000 m surrounded by the high mountains give the possibility of large settlements. Sana'a, one of these settlements, is situated in the geographic centre of Yemen, in the middle of a vast plateau 2,500 m above sea level. The city's climate from May to September is warm and arid during the day and cool and moderately humid at night. From October to April, it is moderate and arid during the day and cold and moderately humid at night. It has large diurnal variations in temperature in both winter and summer. The precipitation occurs mainly in May and August. The wind direction is mainly north-west with an average speed of 2–4.5 m/s. The mean hours of sunshine are from 8 to 11 hours per day during rainy months. Heating during the cold season and penetration of solar radiation to the interior is desirable.

History

It has become fashionable to take an interest in the architecture of the Yemen, the 'Arabia Felix' so called by Roman geographers. On an important trade route of antiquity, inevitably these historic cultures have left their mark, and have influenced recent architecture in Yemen. Around 1000 BC there were three important kingdoms: that of Saba, noteworthy for the biblical account of the Queen of Sheba and Solomon, as well as the temples and dam at Ma'rib, and also the kingdoms of Ma'in and Qataban. History records an influx into the Yemen of Jews, Christians, Ethiopians and Persians. The expansion of Islam brought further contacts as did more recent trade links along the incense route from the far east into Wadi Hadramout.

Derek Matthews,
A commentary on Yemeni traditional architecture.
The British Yemeni Society[4]

Ancient History

Sana'a is one of the oldest cities in the world. Its original name, Azal or Uzal – mentioned in Genesis – is that of one of the sons of Joktan (or Qahtan). Another son of Joktan was Hadramawt, from whom the famous region of Hadramout in South Yemen is said to have taken its name. Joktan was a descendent of Shem (or Sam), the first son of Noah. Ya'rub was another son of Noah, whose grandson was Abd Shams, also called Saba (or Sheba in Hebrew). As scholars are now agreed, Sana'a in ancient South Arabian meant 'well fortified'. The origin of its name probably came from its fortified geography, lying as it does in the great central highland between the high mountain ranges to east and west, and at a point where the plain narrows until it is only a few kilometres wide. This fortified position was undoubtedly chosen to give Sana'a power to control the movement of men and goods from one side of the plateau to the other.

The Kingdom of Saba probably came into existence sometime in the second millennium BC (Lewcock, 1986). Its location in central Yemen enabled it to control the passage of caravans, particularly those of incense – frankincense and myrrh – from the areas where they were produced or imported in the south and east, through to the north. Its wealth derived from the great dam built at Ma'rib, on the trade route which passed along the eastern desert. By the time of the last century before the Christian era, two trade routes existed in Southern Arabia. The oldest and favoured one was that which passed along the desert edge, and was controlled by the Sabaean capital, Ma'rib. The second and less favoured passed over the highland plateau, and thus needed separate control; this was apparently the role of the early town of Sana'a.

After a long history of conflict with the two rival allied states, Qataban to the south and Ma'in to the north of Saba, Qataban became weaker as a result of the emergence of the state of Himyar within it, and Ma'in was destroyed by Saba; this happened some time between 115 and 109 BC. This resulted in the expansion of the Sabaean state to the north and south. It seems that in the last century BC and in the early centuries of the Christian era there existed a greater Sabaean state which owed its cohesion to, and expressed its identity through, the shared worship of the moon god Ilmuqah. The greater Sabaean state expanded its power and territory to include many areas by federation, a number of its rulers being chiefs from the aristocratic clans of the highlands. The status of Sana'a became more important as the Sabaean state expanded in the highlands and before the end of the second century AD Sana'a was established as a second capital, as important to the state as Ma'rib. By the third century AD, the famous high fortified palace, Ghumdan, was constructed and Sana'a became one of the most important cities in Arabia. It is said that Ghumdan was the highest building in ancient times. According to many inscriptions, Sana'a was clearly a centre for royal dynastic families increasingly chosen as the place of residence of the kings of Sabaeans. During that time, the great Sabaean king Yasir Yuhan'im and his sons controlled the whole of South Arabia; they styled themselves 'Kings of Saba, Dhu Raydan, and Hadramawt'. By the fifth century AD, the title was further augmented by adding 'and of their Arabs in the Coastal Plain (Tihamah) and the Highland (Tawd)'. One of the interesting terms given to Sana'a in ancient texts was *'mahram'* which indicates that it had a special status as a royal city, a place to which free access was forbidden or restricted except for royal and high status people, for religious or prestige reasons.

During the fourth century of the Christian era, Judaism began to be observed with the construction of a number of synagogues in Sana'a (Lewcock, 1986). Christianity also made an appearance and a number of churches were built. Judaism remained strong and for another two centuries; the kings of Yemen embraced the Jewish faith and were intolerant of Christians. The most severe of these Himyaritic rulers was Dhu Nuwas ('he of the flowing hair') who ruled from around AD 490 to 525. He massacred the whole Bani Thaleb tribe of Najran, who had embraced Christianity as the result of the preaching of a Syrian. Christian influence was still strong until the year 537 AD when the Axumite general Abrahah seized the Himyarite throne then decided to make Sana'a a site of pilgrimage by the construction of a great pilgrimage cathedral in the city. The arrival of the Himyarite dynasty, based at Zafar, reduced the importance of Sana'a until the Islamic period.

Islamic History

It is believed that Sana'a before Islam in the seventh century was divided into two halves: the first, which was on the east side and considered the older one, was called Al-Qati and included the citadel, the market, the church, and the Ghumdan palace; the second, which was on the west side, was called Al-Sirar and was located between Al-Qati and the Al-Sailah (or 'the rainwater channel' in Arabic) (Marechaux, 1987). Some of the remarkable changes after the arrival of Islam were the construction, in the middle of the seventh century, of the Great Mosque between Al-Qati and Al-Sirar on the site of the Ghumdan palace, and the movement of the *souq* from the periphery of the city to a position next to the mosque, which was a safer area. After that and around the same period, two other main mosques were constructed towards the north side, Farwa Ibn Musaik and Jabbanah. These significant changes, especially the construction of the Great Mosque, had a major impact in changing the orientation of the city's expansion. Thereafter, Sana'a continued to be ruled by governors appointed by the Caliph, the first Umayyad, and after the fall of the Umayyads and the transfer of the Caliphate to Baghdad, Abbasid (Lewcock, 1986).

Muhammad ibn al-Barmaki, the greatest of these governors, undertook a great deal of construction, including a famous underground water supply, or *qanat*, known in Yemen as Ghayl al-Barmaki, which watered the south and the west of the city from a source a few kilometres to the east. With these improvements, the city prospered and expanded. During the Yufirid period in the ninth century, the city was reconstructed and a substantial fortified wall built around it. According to Al-Hamdani's description cited by Lewcock (1986)

> the city continued to grow until a few years after 902 when it was destroyed. Early in the tenth century a survey of Sana'a showed that there were 700 shops, 106 mosques, 12 public baths, 45 sesame mills, and 33 tanning sheds.

With the arrival of Ayyubids between the twelfth and the thirteenth centuries, the city underwent a significant change when the ruler Tughtakin, who was Salah al-Din's brother, incorporated into Sana'a the garden on the western bank of the Sa'ilah, subsequently known as Bustan Al-Sultan. The Ayyubids built their palace there.

Later in the sixteenth century when the Ottomans occupied Yemen the first time, Sana'a was chosen as the administrative capital of the country. They built their ruling palace on the east side of the city, between the Qasr citadel and Sha'ub. In 1597, the governor Hasan Pasha erected the great domed mosque of Al-Bakiriyyah (Lewcock, 1986). In 1679 the Jews were temporarily exiled from Sana'a to Tihama, and then they were permitted to return but not to the old city. After their return, they built a Jewish quarter on the western side of the city which later became known as Qa al-Yahud. In 1872 the Turks occupied Sana'a for the second time and it became the administrative centre of the Yemen until their final withdrawal in 1918. By 1879, the Ottomans had extended the walls of the city to include the farm area to the west side. At that time Manzoni (1884) created a map of Sana'a which has been a valuable resource for historians and researchers ever since (figure

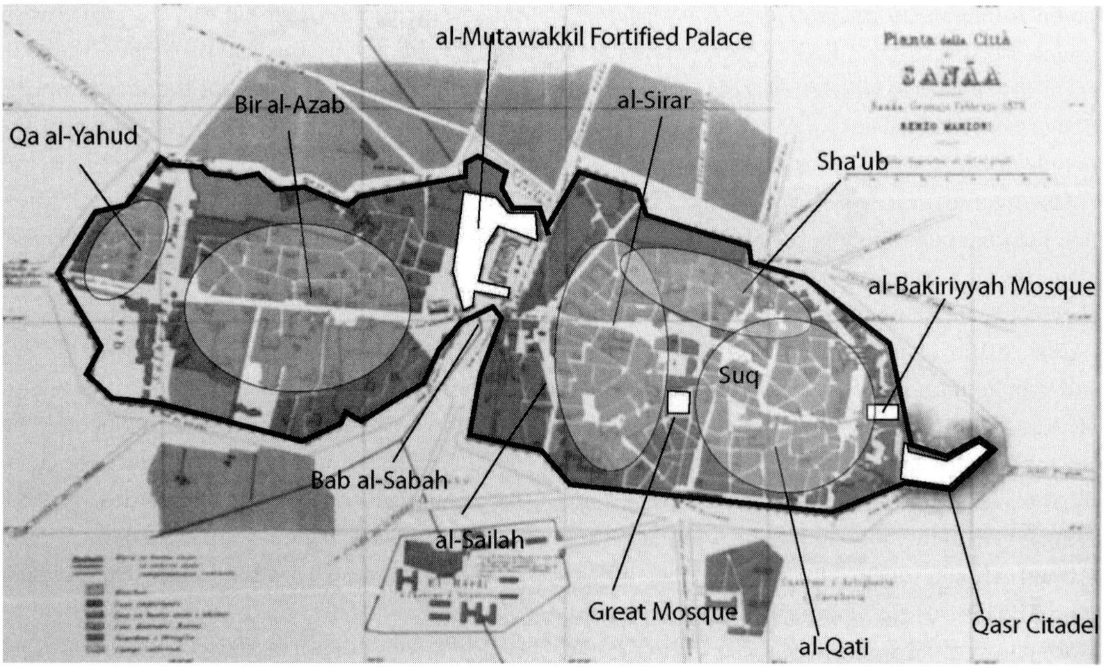

Figure 5.2. Map of Sana'a in 1879 by Manzoni with locations of significant areas and buildings added to it.

5.2). The Manzoni map shows how Sana'a at that time was divided into two halves with a central area connecting them, on which the fortified Al-Mutawakkil Palace was erected. The western part, Bir al-Azab, was developed as a residential area for the *Wali* (the Turkish governor) and most of the senior officers. The city at that time had eight gates and its fortifications were further improved by constructing towers at regular intervals (Lewcock, 1986). The Manzoni map also shows how the *souq* extended from the Great Mosque to Al-Qasr to the east, and to Bab Sha'ub (the gate of Sha'ub) to the north. Other important developments made during this period were the renovation of Al-Bakiriyyah Mosque and the construction of a military academy and two civilian schools. In 1904, the walls of the city were severely damaged after the prolonged siege by tribesmen. The number of inhabitants dropped from 70,000 to 20,000 according to an eye witness, A. Herbert, in 1905 (Lewcock, 1986). Although Sana'a eventually recovered its prosperity and the walls were rebuilt, another intensive siege in 1911 caused widespread damage. The gateway of Bab al-Yemen was rebuilt afterwards with stone and brick.

Modern History

The Period of the Hamid al-Din Imams. In 1904, Yehya Muhammed Hamid al-Din became Imam. After 1919 when the Turks withdrew from Yemen, Imam Yehya built a splendid new palace in Bustan al-Mutawakkil, in which electricity was installed for the first time in the history of Sana'a (Lewcock, 1986.) The Imam made some improvements in the city including various public works. In 1948, Imam Yehya was assassinated and his son Ahmed became Imam. In the same year, the city was sacked by tribesmen. During this attack many buildings suffered damage including the Samsarat Muhammad ibn Hasan which was pillaged. When the Jews left Yemen as a result of the Zionist exodus between 1949 and 1959, their residential quarter, Qa al-Yahud, became a desirable area for many people in Sana'a; it has continued to expand ever since.

Revolution and Civil War: the Establishment of Modern Yemen (the 1960s). The Imams' primitive way of ruling in the north of Yemen isolated the country for many centuries[5] and kept it away from all modern developmental movements in human society. As a result, Yemen suffered greatly from severe political and economic instability causing the people to endure very miserable conditions and poverty. During the Imamic period, the people in Sana'a, as in other parts of Yemen, lived in poverty, with poor education, serious health problems, and in total isolation from the rest of the world; while the people in Aden suffered from the prolonged British military occupation (from 1839 to 1968) and lack of independence. After several failed attempts by revolutionists to change the Imamic political system in the north, in which many revolutionaries were either executed or jailed, at last one planned by a group of officers and led by Abdallah al-Sallal, a high ranking officer in the Imamic army, succeeded, on 26 September 1962, in putting an end to a dark period of Yemen history. The new ideas of the Revolution were born during, and in agreement with, the global movement against colonial occupation and poverty in the Third World. However, the Revolution did not achieve victory easily. Through the 1960s, Yemen was involved in overlapping wars. In the North, a fight between Royalists and Republicans and in the South a fight against the British led to other conflicts among different nationalists. Not until 1970, when the British and Egyptians left and Yemenis found themselves obliged to resolve their disagreements that had evolved through the war, did politics take settled shape (Dresch, 2000.)

Two Yemeni States: Oil Boom, Significant Migrant Labour (the 1970s). When politics took a more settled form at the end of the 1960s, the two Yemeni states found themselves at odds across the 1905–1914 borders. They lived in a state of cold war from about 1970 to 1990, during which each Yemen intermittently denounced the other. While the South formed strong connections with the Eastern bloc, the North did not fit with either East or West. Each built a state system in a period dominated by Gulf and Saudi oil wealth. During the 1970s, there was a boom in Yemeni migrant labour to the Gulf that peaked when the oil states raised prices fourfold after they were provoked by the Arab-Israeli War of 1973. Yemenis did the manual work sending home what in aggregate were enormous sums (Dresch, 2000). The migrant labour's remittances to North Yemen, which had stood at some $40 million in 1969–70, rose to $800 million in 1976–77 and continued rising, to $1.3 billion in 1978–79, dwarfing the revenue of central government. There are no precise figures for the numbers of migrants but a census in 1975 estimated them as 630,000 of the North's total population of 5.3 million at that time, and by the end of the decade the number increased to 800,000. Almost every migrant put his savings into building a house for his wife and children, and the price of building land rose, particularly in Sana'a where on some streets it reached that of European cities. The cost of living there rose fivefold in the 1970s, and the cost of housing about ninefold. The population of Sana'a grew from perhaps 90,000 to 200,000 in the course of a decade. During the civil war period (i.e. the 1960s), a few streets in Sana'a acquired cement buildings of Egyptian type, but the city as a whole remained as in Turkish times. Now all the land around the town that once was fields became building sites, with men cutting stones and pouring concrete.

Yemen in a Wider World: Economic Instability, Cultural Change (the 1980s). The oil boom of the 1970s helped Yemen, through the remittances of labour migrants, to develop and consequently the amount of building construction was enormous, but the effects of declining oil prices in later years were uneven (Dresch, 2000). With the help, from 1984, of locally produced oil and gas, the state apparatus in the North grew faster while the South reached deadlock politically. The global changes in the 1980s were so fast and immense that the context was far beyond Yemeni control. World commodity prices affected the whole area at once, and at the end of the decade the structure of global politics changed dramatically. The difference between the 1970s and the 1980s is remarkable. As Dresch (2000) pointed out:

> A suburb of the *nouveaux riches* emerged near Haddah, just south of Sana'a. A certain symbolic transition had been made, perhaps, when the traditional meeting point between Sana'a and the tribes of Bani Matar became the site of a Kentucky Fried Chicken outlet; now the inner ring-road was well within the built-up area, and building began along an outer ring-road which linked major truck-routes west and south. Haddah to the south and Rawdah to the north became suburbs, while the way in which the whole city functioned changed.

In the late 1980s, supermarkets started to appear in the city and people depended on them. It was a remarkable difference when one remembers the 1967–68 siege when the inhabitants in every house in Sana'a depended on their own wells and what was in their store rooms. The urban construction boom produced a new type of architecture which had very negative impacts on the ecology and water resources (Dresch, 2000). By the early 1980s, the hand-dug wells of 20–30 metres depth had been replaced by bore wells up to 120 metres, and later in the same decade, Sana'a was drawing its water from an aquifer 300–900 metres down resulting in 5–6 metres yearly decrease in the water table. Many large suburban villas, of the kind common in the Gulf, isolated from neighbours by high walls, were constructed in areas such as Haddah. In the past Haddah had been an attractive picnic area with large trees (almond and peach) and dense greenery; now the water was gone and the trees had died.

Yemen as a Single State: Unification Challenges, Sharp Economic Decline (the 1990s). The unification of Yemen was declared at last on 22 May 1990 after a great deal of effort and more than 20 years of disagreement between the two former states. The joy of the unification did not last long when everyone was shocked by Saddam Hussein's invasion of Kuwait on 2 August 1990. Yemen had the great misfortune to be on the UN Security Council at that time – the only Arab state among fifteen members (Dresch, 2000). When Iraq's invasion of Kuwait was formally condemned, Yemen was the one abstention. Yemen spoke for an Arab solution to the crisis, which was then unspecified, in the Security Council. This implied support for Iraq, and on 19 September 1990 the Saudi government revoked the special status of Yemenis within the Kingdom. Hundreds of thousands were forced home after they were obliged to sell at pitiful prices whatever they could not take by truck. On 27 April 1994, fighting started between General Popular Conference of the North and the Yemen Socialism Party of the South. Announcement of secession on 21 May by Al-Bid, the YSP leader, and the obvious support by Saudi government for him provoked the vast majority of people who had been against the war; this resulted in changing public opinion to support for the war against the secessionists. The army found no resistance from the tribes on whom Al-Bid seems to have placed weight. Fighting lasted until Aden fell on 7 July 1994 and Al-Bid and his supporters fled.

The Present Day: Tough Globalization Issues (the 1990s and the New Millennium). The challenge at the present time is how Yemen can survive in the wider world economy. In the last two decades of the twentieth century, migrant labour funded ordinary people's lives; but now the world is shrinking and new forces of globalization have emerged. The changes in global trade and labour require central strategies at a much wider scale. As Dresch (2000) pointed out:

> To live a quiet middle class life in Sana'a at the end of the decade cost about US $1,000 a month. The official salary of a minister was about $US 200 per month. A junior civil servant who held a degree was reckoned to earn YR8,000; rent alone, took YR5,000, utilities, another 1,500, and food for the family about 15,000, or a monthly total of four times his official salary . . . If in recent years some Yemenis have re-established access to the Gulf states and Saudi Arabia, the numbers are insufficient to displace the importance of central government. The economy is an issue as it never was before and in certain respects accords more directly with the world system than with the needs of local farmers or small shopkeepers.

The Vernacular Architecture of Sana'a

> *Viewing the old walled city of Sana'a for the first time creates an unforgettable impression. And this vision of a childhood dream world of fantasy castles is not dispelled even on closer acquaintance.*
>
> Ronald Lewcock, *The Old Walled City of Sana'a*, 1986

Lewcock (1986) gave a detailed description of the architecture of the old city of Sana'a. The predominantly square tower houses impress the visitor with their height. Many houses are more than five levels high, the largest commonly having seven, eight, or even nine levels. A view of the city from a distance, with many hundreds of these houses soaring above the city walls, makes an unforgettable impression. The streets are generally narrow and flanked by towering houses with no sight of vegetation or water to relieve the eye, yet behind the houses and extending right up to them there are frequently large gardens. The urban gardens of the old city of Sana'a occupy one-fifth of the city area; thus almost every house has a view through its windows into extensive gardens.

The urban pattern of the city comprises many housing clusters that are similar in the orientation and spatial organization of their elements, but different in shape and size. It follows a hierarchical system of organization that seemed to be very efficient in providing appropriate functional spaces for both public and private zones and the required separation between them. This system comprises several levels of privacy which start with the most public and ends with the most private. The roads bind the different zones of the city together. They divide the housing zones into quarters and connect them together and with the public zones, such as the market and the large mosque. They are usually narrow, and open onto communal spaces used for gatherings and social events. Each housing quarter includes several mosques and is usually named after the largest or main mosque. The housing quarter is divided into a number of housing clusters, which are grouped around urban spaces. Urban spaces are of two distinctive types (Al-Sabahi, 1996). The first, called *Al-Sarhah* or the social square, is for public use and is usually connected to a mosque and used for people gathering together and for social interaction especially on social or religious occasions. The second, called *Al-Bustan* or the urban garden, is for semi-private use and provides a space for a cluster of neighbouring houses to use as a place for gardening, social interaction, and for children to play. The three main elements that shaped the general urban pattern of the old city of Sana'a are the city roads, the social square, and the urban garden. As seen in figure 5.3, a typical housing cluster is centred on the urban garden (E) and contains several tower houses (C) that surround the urban garden and attached to the social square (D) and the road. The road creates another row of houses, of the same typology, and connects the houses to the social square that includes the mosque (A) and other houses. Figure 5.4 shows façades of typical tower houses seen from the social square and figure 5.5 a view of a typical Sana'a garden.

The Sana'a tower house was usually built for one large family which often includes two or even three generations living together. Family growth within the same house is still a main feature in the Yemeni family. This required designing the house, functionally and structurally, to allow for expansion that usually takes vertical direction to conform to the urban pattern of the city. To achieve this, the house is provided with a staircase which is placed where it can serve all zones on the different floors while not hindering the growth of the house in the future. The walls of the lower floors and the foundations are constructed with thick and strong material to provide the required structural strength for future expansion. The tower house walls are built of 50 cm square black volcanic ashlar stone on the lower levels (i.e. up to approximately 6–10 metres above street level) and 40 cm baked exposed brickwork above that. The roof consists of a frame of wooden beams set 50–60 cm apart, covered by branches and twigs, on top of which lie layers of finely sifted earth, wet and compact, up to a thickness of 30 cm.

The ground floor is frequently double-storey in height and used for keeping animals and wood storage. The first floor is used for grain and food storage. The second floor provides a space to receive guests. On the middle floor there is the

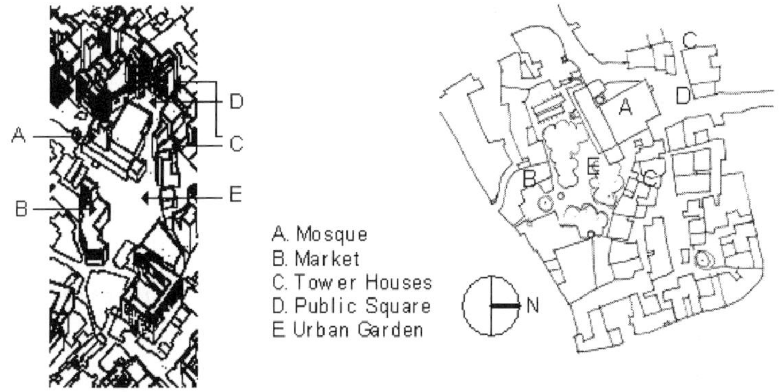

Figure 5.3. Isometric and plan of a typical housing cluster (Masjid al-Shahidayn) in the old city of Sana'a.

Figure 5.4. Tower houses seen from the public square.

Figure 5.5. View of a typical Sana'a garden.

main family room called *Al-Diwan*, which can be used also for social occasions such as parties celebrating a birth, funerals and feasts. The upper floors are residential quarters and include rooms that are functionally polyvalent and non-specific; rooms can be used interchangeably for eating, sleeping, recreation, and domestic tasks. This flexible use of living space is reflected by absence of cumbersome furniture. The living spaces are frequently oriented to the south to benefit from the winter sun and also to escape the summer sun that strikes north façades; a phenomenon that occurs only in the summer time on locations near the equator, such as Sana'a (near the latitude of 15° 31'), where the sun moves towards north of the equator. Sana'a tower houses generally do not experience low altitude sun, coming from east or west orientations, because their east and west sides are usually protected by attachment or proximity to other neighbouring buildings. The top floors usually have rooms to enjoy the panoramic views of the gardens, the city skyline and remote mountains, called Al-Manzar or Al-Mafraj. The kitchen is normally located near the top floor to provide a space for the women in the house to prepare food and have connection to an outdoor courtyard roof for laundry and drying of clothes. To keep the privacy of women using the outdoor courtyard space and prevent them from being seen by strangers in the street or in other houses the roof walls are usually high.

If indigenous Sana'a houses are distinguished from others by their fascinating façades, the unique windows are without doubt their most outstanding elements. The distinction of these windows lies not only in their delightful designs but also, and more importantly, in their different functions. The variation in function from one space to another is reflected in the variations in the size, shape, and number of windows. Generally, such windows have four major parts: the upper section, the lower opening, the vents, and the overhang (figure 5.6). Each part has a different function. One window can consist of some

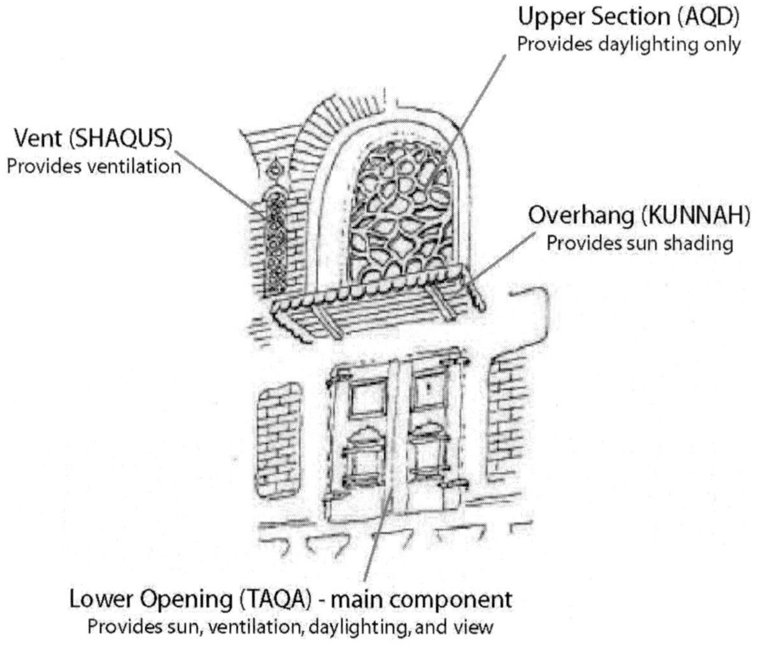

Figure 5.6. Components of the traditional Sana'a window.

or all the parts, depending on its main function and its location on the façade as well as the plan of the house.

Early Transformations in Sana'a after the Revolution

In the first decade of the Republic, the number of transformations that happened to the city was limited; that is due to the civil war and lack of political and economic stability. These transformations emerged mainly as a result of the dramatic change that happened to the political system and the new ideologies that accompanied the Revolution. The early transformations can be explained as follows:

- Demolition of large parts of the old Sana'a walls – these walls were believed by many to be a symbol of the absolute authority of the old regime of Imams and of the total isolation enforced by them; breaking the walls made another sign to people. It represented breaking the prison (or the handcuffs) enforced by the Imams who locked Yemen for centuries. Nevertheless, what caused considerable damage to these walls was the Royalists when they attempted to take over Sana'a again in 1968 by attacking it with heavy artillery during the Seventy Days War with the Republicans. The damage was said to be not perhaps as much as might have been expected considering the capabilities of modern weapons (Lewcock, 1986).
- Creation of the great public square, Square of Liberation (or Maydan al-Tahrir) – the creation of Maydan al-Tahrir next to the ruling palace after demolition of Bab al-Sabah in 1966 was another symbol demonstrating the end of the old regime.
- Building the first tarmacadam roads around the outside of the old city walls – the construction of these roads by the Republican regime was among the first actions towards modernization of the city. They later became the main axes of expansion for the new areas and also the main entries for the new city from other major cities such as Al-Hodeidah and Taiz.
- Emergence of the first modern buildings – these buildings, which appeared first on Maydan al-Tahrir then on other main roads, followed the architectural style known in Egypt in the 1960s[6] and emerged as a reaction to the new ideas of the Revolution that called for modernization and opposed all ideas left from the days of the old regime. They were designed and constructed with the help of Egyptian expertise. They typically comprise shops on ground floors and apartments on upper floors which are used for businesses or housing; and were built by newly introduced materials such as cement (figure 5.7).

Figure 5.7. The first modern buildings to emerge in Sana'a, which were designed and constructed by Egyptian expertise in the late 1960s and early 1970s.

- Early expansion of the city started after the end of the civil war in late 1969 (Lewcock, 1986). It rapidly expanded to the west with the erection of Sana'a University and a residential area around it. On the southern side of the Bir al-Azab wall, new residential suburbs emerged including new embassy buildings. Also, on the northern side of the Bir al-Azab wall, that is, to the south of the Wadi Dahr Road, other residential areas also emerged. The Al-Hodeidah Road and the road to Wadi Dahr were chosen as sites for important new government buildings. Other similar early expansions took place along the road to Al-Rawdah and along the road to Taiz.

  When Sana'a began to grow out of its walls after the Revolution in 1962, certain geographical and political forces played significant roles in determining axes for the city's expansion.

  The connection of the city to other main cities such as Al-Hodeidah and Taiz resulted in forming two main axes; these were Al-Hodeidah Road[7] that connects Sana'a to Al-Hodeidah city and other areas in the west and Taiz Road which connects Sana'a to Taiz, Aden, and other areas in the south.

  The relationship of the city to nearby villages such as Al-Rawda, Haddah, and Wadi Dahr that historically had strong social and trade links with Sana'a resulted in the formation of two important axes: Al-Rawda–Hada Road[8] and Wadi Dahr Road.

  The physical form of the city shaped the expansion axes either as parallel to the outside walls of old Sana'a as in the case of Al-Hodeidah Road or as perpendicular to one of the city walls and connected to a main gateway as in the case of Ali Abdulmoghney Street which is connected to Bab al-Sabah, and Taiz Road which is connected to Bab al-Yemen.

- Construction of the new airport – with the rapid development of the country, Yemen became much more connected with the outside world and so needed a more modern airport to replace the old obsolete one. The construction of the new international airport on a site on the northern side to the west of al-Rawdah was the impetus for the development of the city to the north-west (Lewcock, 1986).

- Emergence of new building typologies – the new regime was eager to create a modern system of administration and to modernize social life. This required the modernization of the entire administrative and service system of the city; new government departments and private organizations were needed to serve all aspects of the developing society. Accordingly, this required modern building types, some of which had never existed before in Sana'a, such as banks, schools, hotels, universities, and factories. The wide acres of land around the city that were previously owned by the Imam were used to construct the new buildings of the Republic or were given to investors who supported the Republic to build new projects. Many of these buildings were constructed on the new main roads, for example the National Bank, Taj Sheba Hotel, and the Army Officers Club.

The Vast Growth of the City

Later in the 1970s and after, the population of Sana'a increased greatly as a result of the large number of migrants who came from other locations in Yemen to the city to enjoy the benefits of job security and the development of the capital. The growth in population increased every year, producing more expansion in the city and more and more development of new areas and buildings. The remittance money coming from migrant labour in Saudi Arabia and the Gulf states played a major role in the prosperity of the construction economy in the 1970s and 1980s. This huge, spontaneous construction movement, which should have been controlled by the government decades before, has resulted in severe urban problems, inappropriate architecture, and serious threats to the old city of Sana'a.

With the new political system that required a large number of staff to serve in the central government, many jobs were created especially during the first two decades after the Revolution (Marchaux, 1987). The benefits offered with these jobs, such as a monthly salary and social security, modern facilities, and a modern life style, attracted people to migrate from their original areas to the capital, most coming from villages and remote areas. According to one estimate, in 1975, 60 per cent of all government employees were in Sana'a and one-third of the total government funds were spent on the development of the city.

After the Revolution and for almost two decades, there were several generations who were poorly educated as a result of the primitive educational system of the Imams. People who lived in Aden or abroad, had better opportunities for modern education and work skills. They were considered very valuable as they had the potential to support the development of the new society. Accordingly, a considerable number of people with modern educational backgrounds or work skills migrated from the more educated areas such as Aden to Sana'a. In fact, some of these migrants were originally northerners who migrated to Aden before 1962 for work because of the poor conditions during the Imamic period. Yemen also attracted a large number of Egyptian and other Arab workers who were much needed for their professional expertise. They helped greatly to develop both the government and private sectors in Yemeni society and to modernize the army and defence capabilities.

In 1934, the total population of Sana'a was said to be18,000, including 5,000 Jews. However, this might be an underestimate (Marchaux, 1987). In 1962, the population reached 55,000, and in 1972, 91,795, of which 68.3 per cent were born in Sana'a and 67 per cent had no formal education. Figure 5.8, based on the 2001 HABITAT report on human settlements shows the projected sharp increase in the population of Sana'a for the period 1985 to 2015 (HABITAT, 2001).

The enormous increase in the city's population has resulted in huge and rapid expansion of the urban and residential areas (figure 5.9). By 1987, the total area of the old city of Sana'a soared from 100 hectares to 2,000 hectares, expanding in all directions outside its walls; that is a twenty times increase in only 25 years. The city's area today is estimated to be around 3,000 hectares; that is thirty times larger than it was 40 years ago. The city has been expanding along some main axes:

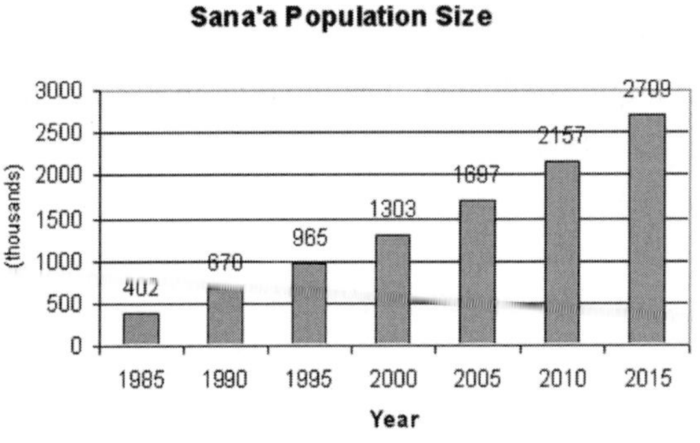

Figure 5.8. Projected increase of population in Sana'a, 1985–2015 based on estimates from HABITAT (2001).

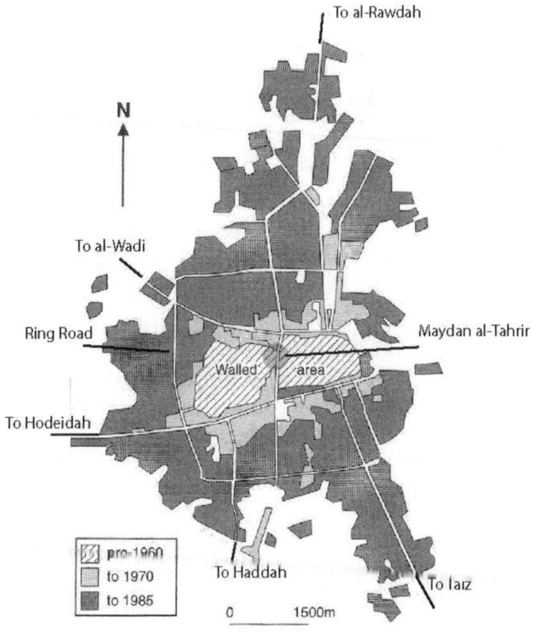

Figure 5.9. Map showing the expansion of Sana'a during 1970–1985 and the main roads and axes of expansion.

- Al-Hodeidah Road, which goes outside the south wall of the city and connects Bab al-Yemen (the Gate of Yemen) with the road to Al-Hodeidah city. This street is now considered a major axis for administrative, banking, and commercial activities as it includes many important governmental and private sector buildings such as the Central Bank, other important banks, the presidential office, and some insurance companies.
- Al-Matar (or Airport) Road, which connects the main street, Ali Abdulmoghney Street, to Al-Rawdah district on the north side where the new international airport is located. Ali Abdulmoghney Street goes from the central axis of the city, the location of Al-Mutwakkil Mosque, the Imam's palace Dar al-Shukr[9] and Bab al-Sabah,[10] to the airport road.
- Ali Abdulmoghney Street, which goes from the south side of Maydan al-Tahrir to Hada residential district. Like Al-Hodeidah Road, this street is another major axis for administrative and commercial activities. In addition to private companies, shops, and dealerships, it includes important banks, the central post office, the National Museum, the Army Command Centre, Taj Sheba Hotel and the Police Officers Club.
- Taiz Road, which connects Bab al-Yemen with the road to Taiz and Aden cities. This street includes mainly low-cost lodging and commercial activities in addition to residential buildings and workshops. It is the main access to Sana'a from major cities and other areas located in the south and south-east such as Dhamar, Ibb, Rada, Al-Baida, Taiz, and Aden.
- Al-Haddah Road, which connects Bab al-Balqa and Al-Solbi area to Haddah residential district in the south. It is the location of many recreational and commercial establishments such as cafés, restaurants, a cinema complex, and all types of shops. It also includes some embassies, an office building complex that includes some governmental administrations, and air travel and tourists agencies.
- Wadi Dhilaa Road, which connects the northern side of the city to Al-Wadi area and to other major cities or villages such as Shibam (to the north), Kawkaban, Thula, and Al-Tawila. It includes several educational areas such as the new main campus of Sana'a University, university academic staff housing, student housing, and on a separate site the College of Medicine campus. It also includes a campus for the Science and Technology University and in a separate site the students' housing.

The Impact of Modernity

The rapid growth of the city should have been accompanied by the construction of infrastructure and services to meet the needs of that growth; however that was beyond the financial and technological capacity of the new state. As a result many random housing and commercial areas have been developed with poor infrastructure and services. Unfortunately, the city that was once perceived as a model of harmony between cultural identity and ecology is now surrounded by new areas that suffer from the predicament of many modern cities, including environmental, cultural, and functional problems.

Ecological Issues

The huge expansion of the city and the change to a modern lifestyle resulted in the consumption of much more water and energy as people installed new household equipment such as washing machines, dish washers, showers, and toilets (Marechaux, 1987). Sana'a inhabitants in the past always relied on underground water by digging wells down to the water levels; water used to be abundant and available at higher levels than today because of the limited use and smaller population. Unfortunately, the huge consumption of water has resulted in greatly lowering the water table and the drying out of many springs. Sana'a today is seriously threatened by water shortages; yet there have been no serious measures to solve the problem. The poor quality of the drainage systems has created a polluted environment which poses serious threats to human health and the city's architecture. Many valuable buildings inside the old city have deteriorated as a result of water leakages caused by the existing bad drainage systems.

Social Issues

People who come from the same social class or have the same tribal roots usually live in the same area or quarter inside the old city of Sana'a. They live together in a homogenous community and develop strong social relations that usually extend over several generations. Over time they build strong relations of trust and cooperation and develop a very strong sense of community. The offer of help by those who can provide it to the other members of the same community is usually regarded as an act of tribute among men; and reluctance to offer help can be seen as shameful. People of the same community share together their happy events such as weddings and feasts, and support each other during sad times such as in sickness or death. The vernacular architecture that was built on the concept of identified urban quarters in which several clusters of houses are arranged with a mosque around social squares (from the public side) or around gardens (from the private side), supported the concept of community living. The social square is a very dynamic element in the urban quarter of old Sana'a, especially if it includes a mosque. It is a place for the people to celebrate religious and social events and to strengthen social relations.

The traditional Sana'a house was built to accommodate a large number of family members, often with three generations living together in the same house. Today, however, new generations usually prefer, at the time of marriage, to leave the family house and live in a separate house or apartment. This is because of the enormous changes in Yemeni society which followed the Revolution. Many young people now see freedom from the restrictions of the extended family house as an important issue. Improvement in the education system has had a major effect in creating self-governing generations who can decide their futures more independently than their parents. Also, after the Revolution, people became much more exposed to outside cultural influences as a result of the improvement in ground transportation systems, the construction of local and regional roads, the construction of international airports, and the foundation of a national airline. Then, following the introduction of public media in the 1970s, such as TV and radio broadcasting and satellite TV and digital information and communication technologies (i.e. computers, networks, the Internet, e-mail, and mobile phones), interaction with other cultures became much wider, easier, and faster.

In the past sons inherited their father's profession, but today most new generations start their careers independently of their family's tradition in a certain profession. This, of course, endangered traditional crafts including those related to architecture and building construction. Another factor that resulted from new generations leaving the family house is the feeling of financial independence. In the past, the grandfather or the father in the traditional house had the upper hand in deciding all financial matters. Adult sons now sometimes cannot accept this and accordingly it can lead to family conflict. The major obstacle that hinders many young people from leaving the family house is the high cost of building or buying a new house. Many who have stayed in Yemen and have not had access to foreign currency incomes (e.g. working in the Gulf countries or supported by relatives there) cannot afford to buy or build a house.

Community Living

Unlike the old city quarters, in which people from the same social class or tribal background live together in a homogenous community and develop strong social ties that usually extend to subsequent generations, those living in the new housing districts come from very different social classes and tribal backgrounds. In the new districts, one can find communities with very heterogeneous social backgrounds living in the same street or even in the same building. Because of this high degree of social heterogeneity, social interaction is limited, especially at the level of older generations. Some people find that the social proximity in the old city intolerable because it allows strangers to interfere in their private affairs; accordingly they prefer to move to the new districts of Sana'a.

Old Sana'a in Crisis

The original inhabitants of old Sana'a abandoned their houses because they could not afford to maintain them, and preferred new villas out of town.[11] The main facilities of the old city of Sana'a such as shops, banking and government shifted out of the old city, mainly to the west and northwest. Educational, recreational, entertainment and health facilities also moved away to an area outside the walls. Most of the wealthy residents of Sana'a moved away due to the unsanitary condition of the streets, lack of services, and the relative inaccessibility of their houses to vehicles. Some of these houses are rented to villagers who come to work in the city or to foreigners who live temporarily in Yemen without their families. The rents for the houses are usually so nominal that they do not help much towards the maintenance required and accordingly the houses are left in a poor condition. If this situation continues, deterioration of these historic houses will increase as will the threat to the social integration of the city.

In 1984, UNESCO and the Yemen government established the 'Campaign for the Preservation of the Old City of Sana'a', with the goal of preserving the architectural heritage of the old city.[12] Ronald Lewcock, an active advocate of the plan, summed up the primary motivation for this campaign:

> Its value lies not so much in the merit of the individual buildings, important though they may be, as in the unforgettable impression made by the whole – an entire city of splendid buildings combining to create an urban effect of extraordinary fascination and beauty. (Lewcock, 1986)

The mission of the campaign is to restore and upgrade the city under the direction of the General Organization for the Preservation of Historic Cities of Yemen (GOPHCY). The campaign is outlined in a UNESCO publication which presents a strategy for conservation. The campaign considered not only the need to preserve and repair buildings but also to revitalize the traditional way of life by creating centres of training for young workers while there are a few remaining masters of old building techniques still alive. In addition to the architectural concerns, the campaign also included social, commercial, educational, and economic projects. Finance is provided by the Yemeni government with the continued help of foreign aid.

In 1988, Sana'a was approved by the World Heritage Committee for inclusion on UNESCO's World Heritage List. However, because of the increasing number of violations in the old city, Yemen received a warning from UNESCO that Sana'a might be excluded from the list if the government does not take serious measures towards correcting these violations. The GOPHCY classified the violations in the old city of Sana'a in several technical categories and levels of severity. There are about 1,800 violations in the old city of Sana'a today, out of which 300 are serious ones.[13] The serious ones include those violations that are visually very obvious or are threatening any part of the old city or its architecture. The use of concrete or aluminium to build a new structure or to modify an existing one is categorized as serious. The GOPHCY today faces several tough challenges that could lead to the failure of the whole project, most important of which are the lack of funds and limited authority given to the organization.[14]

Comparative Analysis of Architectural Transformation

The vernacular architecture and urban design of Sana'a followed certain conventions that were well known and practiced by all its inhabitants. These conventions are history-long experiences that have been developed throughout the life of the city to satisfy the human socio-cultural needs and to respond to the requirements of climate and environment. That people abided by these conventions resulted in a consistent architectural language that gave a very unique identity to Sana'a architecture and preserved it for a long time. The modern architecture in contrast has failed to achieve these values, as can be seen in this analysis.

Urban Pattern

The urban pattern of the old Sana'a is consistent and constrained by a number of elements that make up its composition; these are the roads that are constrained by direction and width, the urban spaces that are constrained by area and proportion, and the building forms that are constrained by orientation and building placement and attachment (figure 5.10). The roads along the east-west axis are dominant in most parts of the historical city of Sana'a. Some important values need to be pointed out:

- Socio-cultural factors have a major effect on the pattern of urban elements, mentioned above. The direction of the main roads are constrained by the locations of significant urban elements such as the *souq* or the Great Mosque.

Figure 5.10. A satellite aerial view of an area near Al-Sailah showing the significant ratio of gardens' areas and other urban spaces in the old city of Sana'a.

- The width of the traditional road and the size of the urban spaces were designed to allow (human) functional and socio-cultural activities to be performed comfortably.
- Environmental factors also have a great effect on the pattern of urban elements. A previous study by Al-Sallal (1996*b*) showed that the road directions in the old city of Sana'a maximize solar access to buildings because the roads run mostly east–west which allows the sun to strike south façades.
- The size of the roads were designed to permit appropriate solar access to the housing sections located along their north sides. The narrow roads in some parts were found to hinder the solar access for some buildings. The same study showed that, for areas dominated by four-storey buildings, the roads that run east–west should not be less than 6 m wide and the roads that run north-east–south-west, or north-west–south-east, should not be less than 9 m wide. These sizes were found to be typical in the old city; however, taller buildings required roads of greater width.
- Houses were placed on the roads in a way that maximizes exposure to south sun that is required for passive solar heating. Buildings that are on the south side of the road have solar access from the back gardens. This is an effect of the climate on the city space.
- In many cases, the houses are attached or very near to other houses on the east and west sides which helps to avoid the low angle sun coming from east and west. This is an effect of the climate on the city form.

It is very difficult to recognize dominant or consistent urban patterns in the architecture of the new areas of Sana'a. The layouts in the new districts are so varied and often do not follow any urban rules or standards (figure 5.11). The absence of any policy with regard to city planning and building codes, the lack of urban planning experts, and the lack of control over the widespread construction of housing have been the major causes of the existing lack of uniformity. The majority of the houses built in the new districts of Sana'a in the last three decades were constructed on land which at the time was not controlled by the urban planning agencies of the government. These houses were usually constructed by one of three categories of people: those who migrated from villages to Sana'a looking for work opportunities and better city amenities; those who were labour migrants in the Gulf and probably residents of other areas in Yemen but came to reside in the capital for its improved amenities; and, although in smaller numbers, those who were from old Sana'a but left their original vernacular houses because of socio-

Figure 5.11. Two views of modern roads in Sana'a. Urban patterns are very varied and minimum consideration is given to human spaces.

economic changes and the difficulty of maintaining their original houses. The people of these three categories constructed huge numbers of houses on land they bought from tribal areas around Sana'a either before the existence of any urban planning maps prepared by the government or, in the best cases, with very early road maps.

Housing Cluster

Some important points in the design of the traditional housing cluster need to be emphasized:

- Privacy in the traditional housing cluster was ensured by designing indirect entrances and minimum area of openings in the ground floor (see figures 5.4 and 5.5 above). This is an effect of the outdoor space function on building form. In many instances, the social square is directly connected to a mosque that is needed for the daily prayers as well as to support certain events happening in the social square such as celebrations, feasts, and weddings. The space of the social square was designed to permit the gathering of many people, usually males, to celebrate religious feasts and social ceremonies; and the space of the urban garden was designed for cultivation activities.
- Buildings that were most demanding of solar access, such as houses, were placed along the north–north-east boundary of the urban garden or social square with minimum heights of four floors (figure 5.12). Buildings that were least demanding of solar access, such as the marketplace, were placed along the south–south-west boundary of the urban garden or social square with maximum heights of two floors.

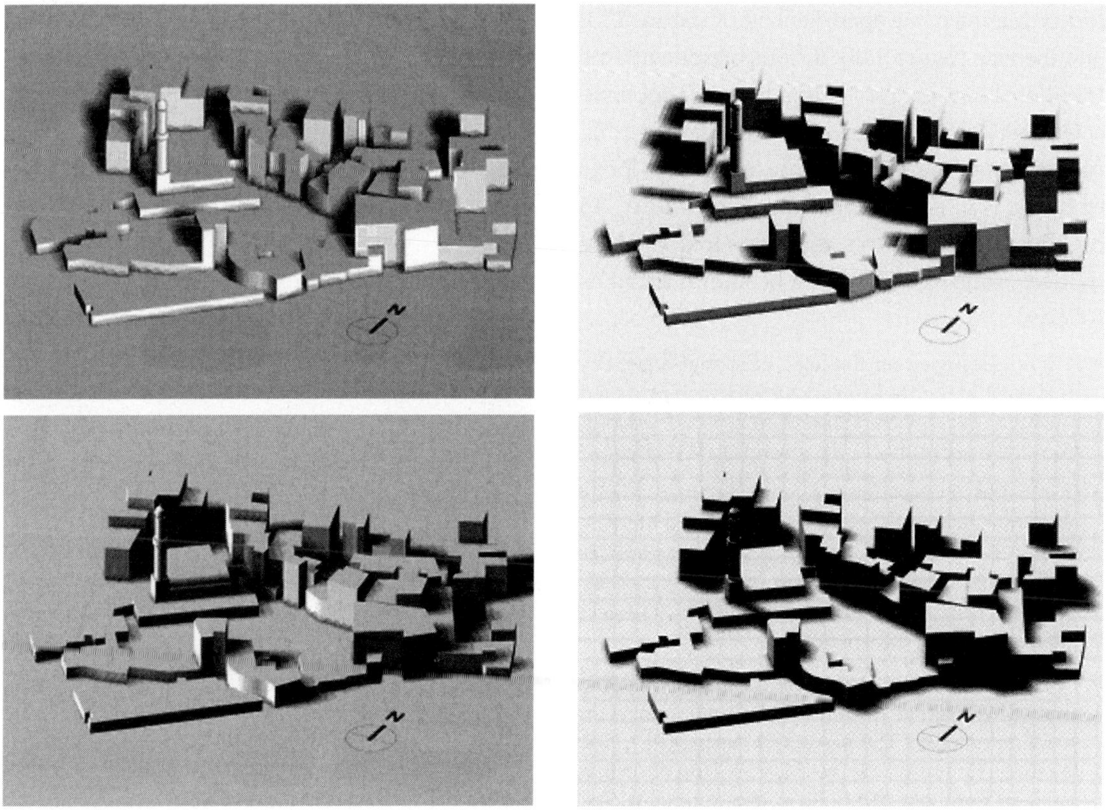

Figure 5.12. Visualizing solar access and shading in a housing cluster case study (Masjid al-Shahidayn cluster) in the old city of Sana'a produced by three-dimensional computer modelling.

By this arrangement, solar access was ensured to the buildings that needed it the most such as the houses. These findings are based on an analysis study which employed three-dimensional computer modelling (Al-Sallal, 1996*b*).

- The size and shape of the urban garden (or the social square) was designed to permit appropriate solar access to the housing sections located around its boundaries. This was achieved because the length of the urban garden (or the social square) in north–south direction exceeded the minimum required distance, 6 m, as recommended by Al-Sallal (1996*b*).
- Another important consideration was the proportion of east-west to north-south dimensions of the urban garden (or the social square). It was found to be 1.66: 1 and according to Al-Sallal (1996*b*) such a proportion has the potential to maximize south solar access to the largest number of housing buildings.

Within urban patterns that characterize most of the new districts of Sana'a, generally two housing forms dominate: the apartment block (figure 5.13) and the villa (figure 5.14). In comparison with the traditional housing cluster that follows a sophisticated hierarchical system with multi-level spatial organization, the modern housing cluster looks so naive as it lacks several important values. To highlight the differences, the shortcomings of the modern housing cluster can be summarized as follows:

- It is not designed on the logic of spatial hierarchy as is used in the traditional cluster which provided multiple levels with different degrees of privacy, from the most public to the most private levels.
- It is based on very simple geometry with highly repetitive elements. Unlike the traditional housing cluster that provided a dynamic visual experience with its rich and changing geometry, the modern housing cluster creates a monotonous and less interesting visual experience.
- Unlike the social square in the traditional housing cluster, it does not provide communal spaces for social interaction. Therefore it forces people to use the roads for social activities instead. This, of course, can lead to unsafe and polluted environments. The lack of such public spaces can also restrict opportunities for solar access in the winter.
- Unlike the urban garden in the traditional housing cluster, the modern cluster does not provide communal green spaces. Therefore it restricts social interaction among neighbours and restricts opportunities for solar access in the winter.

Figure 5.13. Transformation of the apartment block housing type in Sana'a in three different decades (1970s, 1980s, 1990s).

An interesting analysis made by Al-Sabahi (1996) compared the traditional and modern sectors in Sana'a with regard to land use using several indicators: total area of cluster; green space; service area; road and social square space; and population size. As can be seen in table 5.1, these indicators show that the vernacular housing cluster provides more space for social activities and greenery than the modern housing. The same study made a nice visual comparison between the traditional and modern sectors in Sana'a with regard to architectural expression using several indicators: the proportion of width to height; the dynamic nature of the sky line; the dynamic nature of the building line; the type of openings; the ratio of void to solid; and the projecting masonry on the elevation.

Figure 5.14. Transformation of the single family housing type (or villa type) in Sana'a through three different decades (1970s, 1980s, 1990s).

Tower House Design

A southern room is worth one room, a western or eastern room is worth half room, and a northern room is worth no room.

Sana'ani proverb

Sana'a tower houses are surrounded by beautiful views of green gardens and a skyline of the decorated towers' façades and minarets, all with a background of remote mountains and blue sky (figure 5.15). These fascinating views, combined with the mild climate, encouraged the opening of Sana'a buildings to the outside, while those of other parts of the Islamic world open to inside courts. The traditional tower house was constructed according to certain conventions; some are stringent, to respond to the socio-cultural and environmental needs. Its design is constrained by a number of rules that specify its form proportion, orientation, zoning, and space arrangement. Its vertical tower form succeeded in satisfying some necessary socio-cultural requirements in Sana'a (Al-Sallal, 2001). Its slender proportion has the potential to provide highly social connection with the close community around it while maintaining high privacy and safety for its living top

Table 5.1. Comparison between the traditional and modern sectors in Sana'a.

	Al-Abhar quarter (as ratio of the quarter area)	*Al-Madinah al-Sakaniyah complex* (as ratio of the complex area)	Remarks
Area of cluster	30%	28%	
Green area	42% communal	32% private	No communal green area in Al-Madinah al-Sakaniyah complex
Service area	10%	9%	
Roads/social square	18% social square + roads	31% roads only	No communal social square in Al-Madinah al-Sakaniyah complex
Population	525 inhabitants	120 inhabitants	

floors. Another important quality found in the slender vertical form is its potential to provide most indoor spaces with views to the outside in several directions, while still keeping the required privacy. These two important qualities are unfortunately missing in the modern house since it is usually constructed with high solid fences, totally isolating it from its social context.

The vertical form of the traditional tower house also provides other significant bioclimatic advantages over the typical horizontal form of modern house of equal volume and floor area (Al-Sallal, 2001). It helps to improve solar access to the upper floors that usually include the best living spaces. With its large external wall area it allows high exposure to the low winter sun while it restricts exposure to the high summer sun by its limited roof area. The horizontal form of the modern house does quite the opposite. Further, the vertical form of the Sana'a traditional house improves the convective airflow effect by its vertical proportion, and employs it to heating the living spaces located in the higher floors while cooling storage and service spaces located in the lower floors.

A study by Al-Sallal and Cook (1992) concluded that the people in Sana'a found that the integration of the different functions of a window could be performed more efficiently by separating the window into components, each of which was responsible for one or two main functions. This, in turn, provided more flexibility in control-

Figure 5.15. Tower houses seen from the urban garden.

ling each function. The variation in the horizontal arrangement of windows (shapes, sizes, and number) is a result of climatic influences such as solar motion and wind direction. South façades have larger windows with more components and controls than north façades so as to exploit solar energy for heating. The variation in the vertical arrangement is a result of socio-cultural influences such as the people's activities and traditions. While higher floors have large windows to maximize visual connection to outside views, lower floors have smaller windows for privacy.

The window designs found in modern houses, unlike the traditional ones, are designed without any rules or conventions and show no logical order (figure 5.16). In most cases, the designs do not reflect climatic or cultural needs. Also, their variation within the same building does not make any sense either with regard to outside factors such as climate or view or inside factors such as their arrangement in the indoor spaces. While windows were the most beautiful elements in the traditional house of Sana'a, now they have lost their unique architectural identity.

Conclusion: The Need for a Novel Approach

Sana'a today lives in two disconnected, yet

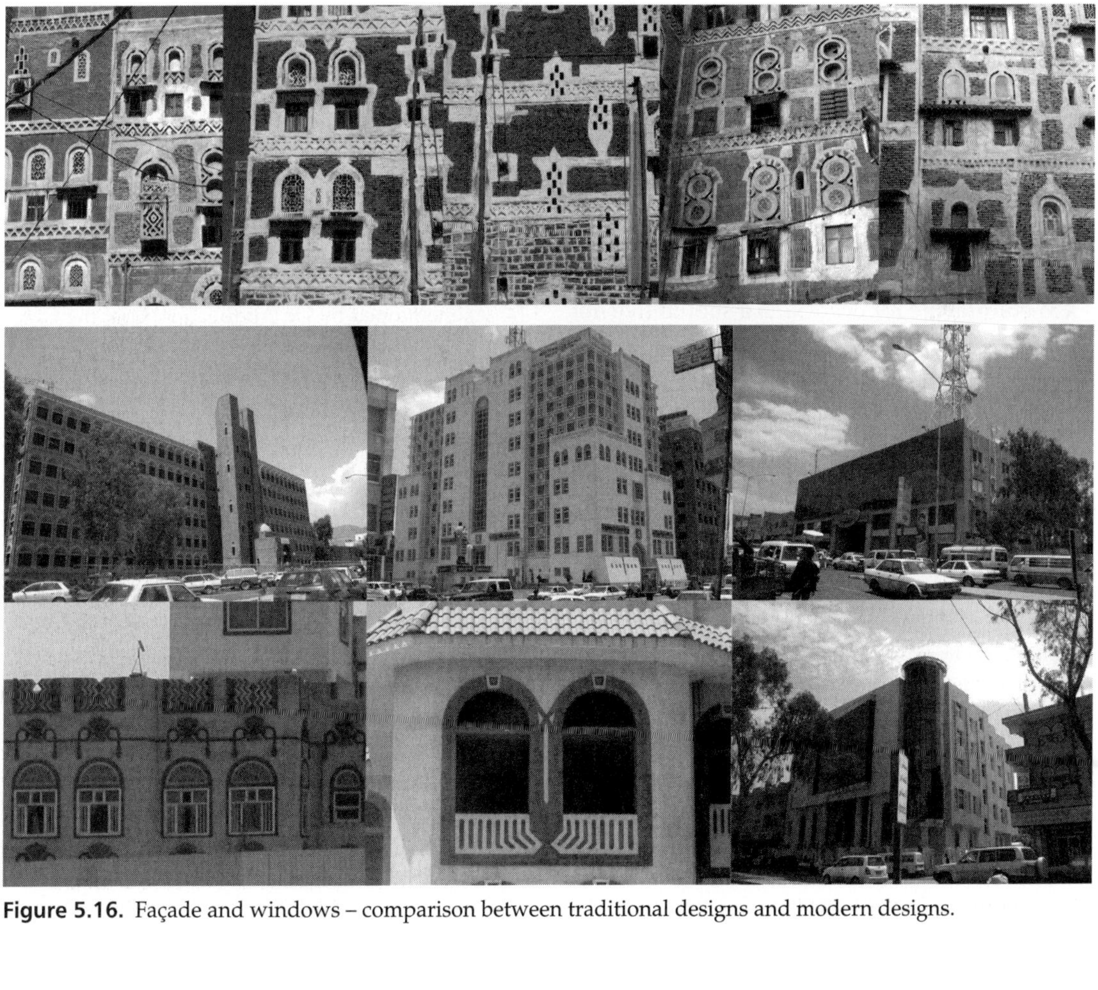

Figure 5.16. Façade and windows – comparison between traditional designs and modern designs.

adjacent, worlds. People of the first look for independence and high standards of modern life and dislike the old unclean life of the old city, while people of the second are proud of their heritage, appreciate community living, and refuse to accept the intrusion of a Western lifestyle. There is no doubt that the old city of Sana'a and its unique architecture should be safeguarded; it is at serious risk today and it seems that the future will probably be worse. The number of violations is increasing every year without any serious measures to stop them. Yemen received a warning from UNESCO that Sana'a might be excluded from the UNESCO's World Heritage List if the government does not take serious measures towards correcting these violations. The GOPHCY today faces several tough challenges that could lead to failure of the whole project, the most important of which are the lack of funds and limited authority given to the organization.[14] It is true that we should safeguard the old city of Sana'a including its heritage and cultural values, yet it is unrealistic to imagine that people will continue to live the same traditional socio-cultural life for ever. Sana'a should live not only because it is a landmark of the great history of Yemen, but also as a resource to teach cultural and ecological values that can help us to create an appropriate architecture for Yemen today.

NOTES

1. The http address is www.new7wonders.com.

2. Quotation from an article '"Sana'a: Pearl of Arabia" seeks global fame' on the website http://www.arabia.com/life/article/english/0,11827,49131,00.html.

3. Quotation from an article titled 'Conservation of the Old Walled City of Sana'a, Republic of Yemen', by T. Luke Young on the website http://web.mit.edu/akpia/www/AKPsite/4.239/sanaa/yemen.html.

4. Quotation from an article in a website titled 'A commentary on Yemeni traditional architecture', by Derek Mathews on the website http://www.al-bab.com.

5. The Zaydi Imams' rule may be summarized as follows: first control of Sana'a 1323–1547; limited control during the first Ottoman occupation, 1547–1630; full control 1630–1872; limited control during the second Ottoman occupation,1872–1918; full and steady control 1918–1962.

6. Which is a continuation of the international style movement

7. Known also as Al-Zubairy Street.

8. Known also as the Airport Road.

9. Known now as the National Museum.

10. This entire area is known now as Maydan al-Tahrir.

11. Quotation from an article titled 'Conservation of the Old Walled City of Sana'a, Republic of Yemen', by T. Luke Young on the website http://web.mit.edu/akpia/www/AKPsite/4.239/sanaa/yemen.html.

12. From an article titled 'Architecture, Modernity, and Preservation: The Tower House of Sana'a, Yemen, by Richard Brooks Jeffery in Aridlands Newsletter produced by University of Arizona.(http://ag.arizona.edu/oals/ALN/ALNhome.html.

13. Information given in by Abd al-Hakim al-Sayaghy, Manager of Public Affairs in the General Organization for the Preservation of Historic Cities of Yemen (GOPHCY) on 18 August 2003.

14. As mentioned by Dr. Abdullah Z. Eisa, the Chair of the General Organization for the Preservation of Historic Cities of Yemen (GOPHCY) on 17 August 2003.

REFERENCES

Al-Oulfi, M. (2000) Characteristics of Yemeni Architecture: Forms and Directions of Development. Unpublished PhD Thesis, Shanghai Tongji University, China.

Al-Sabahi, H. (1996*a*) *Sana'a Yemen Tradition and Modernity in Sanani Architecture*. Sana'a: Sana'a Modern Offset Printing Press.

Al-Sallal, K.A. (1993) The role of indigenous fenestration to reduce energy requirements in the Sana'a house, in *Proceedings of Fourth Arab International Solar Energy Conference*, Vol. 2. Amman: Royal Scientific Society, Amman, pp. 875–888.

Al-Sallal, K.A. (1996*a*) Traditional methods in new

forms: insight to achieve energy conservation in the modern Sana'a house, in Sayigh, A. (ed.) *Proceedings of the World Renewable Energy Congress*, Vol. 3. Oxford: Pergamon-Elsevier, pp. 1438–1441

Al-Sallal, K.A. (1996*b*) Solar access/shading and building form: geometrical study of the traditional housing cluster in Sana'a, in Sayigh A. (ed.) *Proceedings of the World Renewable Energy Congress*, Vol. 3. Oxford: Pergamon-Elsevier, pp. 331–334.

Al-Sallal, K.A. (2001) The balanced synthesis of form and space in the vernacular house of Sana'a: bioclimatic and functional analysis. *Architectural Science Review*, **44**(4), pp. 419–428.

Al-Sallal, K.A., Ayssa, A.Z., and Al-Sabahi, H.A. (1995) Thermal performance and energy analysis for Sana'a vernacular house, in Ben Ghadi S. (ed.) *Proceedings of Applications on Renewable Energy in Yemen Workshop*. Aden University, Yemen.

Al-Sallal, K.A., and Cook, J. (1992) Sana'a historical windows: integration between comfort and aesthetics, in Burley S. and Arden M. E. (eds.) *Proceedings of 17th National Passive Conference of American Solar Energy Society*, Vol. 17. Boulder, CO: American Solar Energy Society, pp. 197–202.

Ayssa, A.Z. (1990) The impact of the traditional window shutters and attachment as comfort producing procedures in the vernacular houses of the old city of Sana'a. *Proceedings of the World Renewable Energy Congress*, Reading, UK.

Ayssa, A.Z. (1994) The vernacular tower house of the old city of Sana'a, Yemen – an investigation of its internal air movement, in Sayigh A. (ed.) *Proceedings of the World Renewable Energy Congress*. Oxford: Pergamon-Elsevier.

Ayssa, A. Z. (1996) A vernacular house in Zabeed, in Sayigh A. (ed.) *Proceedings of the World Renewable Energy Congress*, Vol. 3. Oxford: Pergamon-Elsevier, pp. 1425–1428.

Bahman, Kia and Craddock Williams, Vivian (1989) Saving Sana'a. *Geographical Magazine*, May, pp. 32–36.

Dresch, P. (2000) *A History of Modern Yemen*. Cambridge: Cambridge University Press.

HABITAT (2001) *Cities in A Globalizing World, Global Report on Human Settlement*. Nairobi: United Nations Centre for Human Settlements.

Hirschi, S. and Hirschi, M. (1983) *L'architecture au Yemen du Nord*. Paris: Berger-Levrault, p. 250.

Ibn Rustah, Abu Ali Ahmad ibn Umar, in de Goeje, H.J. (ed.) (1892) *Kitab al-Alaq al-nafisah*. Leiden: Liber viarum et regnorum. BGA, Vol. VII.

Lane, M. B. (1988) *Sana'a: Pilot Restoration Projects*. UNDP/UNESCO YEM/88/006.

Lewcock, R. (1986) *The Old Walled City of Sana'a*. Paris: UNESCO.

Manzoni, R. (1884) *El Yemen*. Rome.

Marechaux, P. (ed.) (1987) *Sana'a: Parcours d'une cité d'Arabie*. Paris: Institut Du Monde Arabe, p. 21.

Matthews, D. (1979) Building in the Yemen. Report in the Proceedings of a Seminar Held at the Museum of Ethnography, Rotterdam, p. 10.

Rogers, H. (1986) Sana'a in the twentieth century. *The Architect*, January, p. 23.

Serjeant R. B. and Lewcock, R. (eds.) (1983) *Sana'a An Arabian Islamic City*. London: The World of Islam Festival Trust.

Veranda, F. (1987) *Art of Building in Yemen*. London: Art and Archaeology Research Papers, p. 83.

Watson, D. (1983) *Climatic Building Design*. New York: McGraw-Hill Book Company.

Chapter 6

Lake Tunis, or the Concept of the Third Centre

Bechir Kenzari

Autour d'elle [Tunis], trois lacs qui, sous le dur soleil d'Orient, brillent comme des plaines d'acier. Au nord, au loin, la Sebkra-er-Bouan; à l'ouest, la Sebkra-Seldjoum, aperçue par-dessus la ville; au sud, le grand lac Dahira ou lac de Tunis; puis, en remontant vers le nord, la mer, le golfe profond, pareil lui-même à un lac dans son cadre éloigné de montagnes.

Guy de Maupassant, *Récits de voyage*,1890

Any standard reading of major cities in the Middle East and North Africa would identify three cardinal urban fabrics: the old city, the modern city, and everything built in the suburbs, whether spontaneously or otherwise. An updated reading of the City of Tunis would also point to another, corresponding to the emergence of the Lake Tunis Project. This city within the city has some distinctive characteristics: (1) because of its geographic location and functioning, it constitutes a third centre alongside the *medina* and the European city; (2) its waterfront of reclaimed land potentially offers unlimited new urban possibilities; (3) it is becoming the alternative seat of tertiary and diplomatic activities;[1] (4) it is close to Carthage International airport, the ports of La Goulette and Radès, Le Kram Trade Fair Center and is easily reached by rapid expressways; (5) it has been designed with the intent of improving the environmental conditions of the city of Tunis as a whole.

This chapter explores this new urban reality in light of the history of Tunis as well as the newly emerging global condition, especially with regard to ecological consciousness in the Mediterranean basin.

The *Medina*: the First Form of Centrality

Since medieval times Tunis has ranked among the major capital cities of the Mediterranean, especially under the Hafside rule between the thirteenth and fifteenth centuries. In the early nineteenth century, it surfaced again as a prosperous city, particularly under the government of Hamouda Pasha (1782–1814). Characterized by an elaborate urban system, organized so as to stress intimacy through its lanes and its patio architecture, it tirelessly sought perfection in the way it organized itself. Integrating Andalusian, Turkish and Italian influences, from the nineteenth century it became a city of renown visited by European tourists.[2] In 1860, the process which led to colonial dependence was well advanced. The capitalist and industrial incursions of Europe, exhaustion of public funds, an oppressive fiscal system, and the Tunisian state's indebtedness had created the conditions leading to the 1864 uprising led by Ali Ben Ghdahem El-Mejri and eventually to the French intervention in 1881 under the device of the 'Protectorate' (Amodei, 1984).

The medina of Tunis was the first to suffer from this colonial hegemony. Originally the medieval city, its *casbah* and the two *rbat* (suburbs) were surrounded by a *sur* (rampart), thus providing the city with a line of defence. The ramparts surrounding the suburbs on the north, south and west sides converge on the east close to the *khandaq* (major sewer lines) thus completing the line of defence to the lake. Under the French administration, the old city proper was surrounded by a ring road that follows precisely the line of earlier city walls (only scattered sections of the wall and a few of the *bab*, or gates, now survive). Within this loop lies a spatial and structural order attuned to the way of life of medieval and early modern Tunis, to the Tunis of the Hafsids and the early Husaynids, famous for its urban landmarks such as the Zitouna Mosque, several *madressa* (religious schools), the *souq* (covered bazaars), lively streets and *zanka* (lanes), and the remaining city gates and ramparts.

With the establishment of the European city, the medina ceased to be the centre of the capital.[3] Built on the low-lying swampy flood plain near the lagoon, which had been rejected by the Arab-Moslem founders of Tunis for reasons of land quality and sanitation, the European city developed along the axis that links the Moslem city to the sea. The notable constituent in this plain is the Promenade de la Marine: a straight road, planted with triple rows of trees, leading from Bab Bhar (the Port of the Sea), to the lake, to a small dock surrounded by the buildings of the Navy and Customs. It was to become one of the most important routes of the future metropolis, as it linked it directly to the lake and beyond through the port canal of la Goulette to the Gulf of Tunis and the Mediterranean Sea. The layout of perpendicular arteries meeting at Jules Ferry Avenue formed the regular design which would influence later urban growth outside the confines of the Arab city. Despite the construction of numerous buildings which shaped colonial initiatives (the municipal theatre, central market, cathedral, Lycée Carnot, arcades, sumptuous buildings with highly floral façade treatments by Sicilian masters), the native population was left aside and therefore identified with the old city, culturally, politically and urbanistically. The medina itself, somewhat meaningfully, was bordered by commanding colonial edifices (barracks in the Casbah, law courts in Bab Benat, Charles Nicolle Hospital and the central jail in Rabta). The old city found itself strangled from all sides by the symbols of French might, the menacing signs of power.

It is true to say that during French rule an urban crisis was simmering. The crisis was manifest in the lack of new urban development and buildings by the colonial state for the local population. And it was a unique phenomenon in the capital that all the colonial buildings that emerged on the periphery of the medina were themselves surrounded by spontaneous settlements inhabited by the poor local Moslem Tunisians. Thus Montfleury was surrounded by Saida Manoubia, Bellevue by Borj-Ali-Rais, Rabta and Tawfik by Mellasine, Franceville, and

Mutuelleville by Jebel Lahmar. Colonialists and locals remained segregated in different areas, in an alternating fashion: natives, colonialists, then natives again. The effects of this urban segregation will be seen later in relation to the first explosion of spontaneous housing.

At the time of Independence in 1956, the medina began to witness the phenomenon known in French as *taudification*, that is its conversion into a slummy, decaying area. With the considerable exodus of the upper- and middle-class bourgeoisie from the medina to the modern city, a corresponding exodus took place whereby the houses in the medina were occupied by migrants from the countryside. Between 1956 and 1972 the old city and the two *rbat*, an area of 270 hectares, became the main reception area for these new urban dwellers. Today 65 per cent of households living in the medina and 50 per cent of those in the northern and southern extensions are migrants; some 40 per cent of the migrant households settled in Tunis during the first decade of Independence. Densities are very high (550 inhabitants per hectare). Pressure of numbers has resulted in the subdivision of many traditional houses, originally single-family residences but now occupied by several families. There are some 15,000 houses in the medina and the *rbat*, occupied by 30,000 families; 45 per cent of all families live in one room each. In the process of subdivision, the nature and significance of the houses also changed. When the house was occupied by several families the patio, originally a private space, became a public space (in practice an extension to the street). Severe overcrowding, subdivision among poor families possessing few resources, and high rates of tenancy (over three-quarters of households live in rented accommodation) have contributed to the rapid physical deterioration of the city's traditional fabric through lack of maintenance. Housing conditions are therefore extremely bad; half the dwellings are in a poor state of repair, and some lie in ruins.

Thus transformed, the medina has become the most prominent low-income commercial centre in the whole city. New commercial activities have expanded considerably (particularly textiles and foodstuffs), responding to an increasing popular demand, and if some long-established trades and crafts have weakened, others have continued to exist and found new markets among the rising number of foreign tourists who visit the medina every year. Traditional commercial activity, partly tailored to tourism, remains around the Great Zitouna Mosque. New tourist shops have expanded along the Rue Djamaa Zitouna while new popular commercial activities have developed along the north-south axis. There are now some 6,000 shops and workshops in the medina enjoying an annual revenue of hundreds of millions of dollars and employing about 16,000 people. The physical expansion of commercial activities has been achieved by conversion of residential and public buildings into commercial spaces, and productive activities have penetrated all parts of the urban fabric. Maintenance of the premises used for commerce and small industries is minimal, aggravating the general deterioration of the traditional fabric, and maintenance of most of the religious, cultural and educational buildings in the medina is also poor. Since the sale of *habous* properties (religious foundations to manage donated properties) many of these monuments have been neglected and the majority converted to new uses – housing, storage and workshops.

The effort needed to implement a *degourbification* policy – the removal of *gourbi* or shacks, i.e. slum clearance – has been presented by the government as a means of rekindling national self-esteem. That is how, immediately after the

Independence in 1956, a modernist vision of spatial organization led to the destruction of some 20 hectares of the old fabric in order to facilitate vehicular traffic. This process of demolition, supported by the state itself, was eventually aborted by the municipality of Tunis through the creation of the ASM (*Association de la Sauvegarde de la Medina*) in 1967.

From 1970 the position of the government changed as it accepted the concept of safeguarding the medina and the integration of the historic area within the general scheme of modernization of the capital. Ever since, and sustained by UNESCO, the ASM has fought to change the governmental attitude towards the old city. The interventions of the ASM operate at two levels: overall policy and specific operations. Firstly, regarding the historic fabric, an overall policy of restoration and preservation was adopted, pertaining to the improvement of the residential, economic and cultural sectors inside the old city. The aim is to stop the destruction of the old fabric and to assist, technically, private individuals in their efforts to improve the condition of their properties. Despite encouraging results, progress has been relatively slow. Secondly, the ASM has sought to create formulae that may help inject new ideas into the old fabric. Al-Hafsia project is a case in point. Judged to be an unhygienic quarter since 1928, this area covering 6 hectares, and later extended to 13.5 hectares, was scheduled to be re-built in three phases, starting in 1977. The implementation of this renewal project, assisted by the World Bank, has been judged a successful operation that has achieved the integration of the new project within its traditional environment. Opinions differ, however, when it comes to another renewal operation, namely Bab-Souika, implemented in the mid-1980s. Covering an area of 30 hectares, this prestigious and expensive project put the emphasis on the design of public plazas, and of a traffic tunnel, rather than on the urban design and restructuring that would benefit the inhabitants of the neighbourhood.

European City: the Second Form of Centrality

The first railways and other modern services had been established before the coming of colonial rule in 1881. Before the Protectorate, the following railway lines were established: from Tunis to Bardo (1872), linking the capital to the small fortified city, a residence with government functions of the Husseinite Beys since the eighteenth century; from Tunis to la Goulette and to La Marsa (1871–1876), along the northern shore of the lake, reaching the English (later Italian) station, to the north of Promenade de la Marine, and branching towards the dock on the lake; the Tunis to Algiers line (1876–1880) through the Manoubia Tunnel to the south, followed by the Hammam Lif Line (1882), reaching the 'French' station south of the Promenade, also branching towards the Port. Among the developments the following can be mentioned: the new Consulate of France, built by engineer Colin according to a design by Architect Caillat begun in 1862 on Promendade de la Marine; the Central Market, in the southern quadrant near the old city; the Gas Factory for public street lighting, belonging to an English company (1874), built in the southern quadrant towards the lake (Amodei, 1984). The Port of Tunis was designed and finished in 1893 by the Batignolles firm in order to facilitate the navigation and docking of ships coming from the Mediterranean via La Goulette and the Tunis Canal.

From 1881 Tunis was confirmed as capital of the Protectorate, connecting the principal political and economic decisions of the metropolis and the country and serving as the distribution centre for imported goods. It also became a bipartite city. Alongside the historic city, weakened in its key roles and where almost all the Tunisian population lived, was the modern city, built on a Western pattern, inhabited by a foreign population, and connected with the colonial government. It was here that the services and facilities (banks, commerce, industry and transport, head-offices and stores) were rapidly installed. Formed around the axis that stretches from Bab-Al-Bahr (Sea Gate) to the Port of Tunis the European city developed on relatively low-lying reclaimed land with a regular layout of perpendicular, hierarchically organized avenues, streets and passages. Because it was low-lying, it tended to be frequently flooded by heavy rains. Despite this, the visitor can find pleasure in exploring its streets and avenues and the diverse architecture of this essentially twentieth-century capital.

Under French rule, Tunisian Moslems and Jews lived in the medina while Europeans (Italians, generally from Sicily, Protestant Maltese, Jews of Livorno[4] and, of course, the French) lived in the modern city. There was also a Greek and a Russian minority. These communities had their respective languages, places of worship, their social circles and their newspapers. The building sector, from design to construction was an Italian monopoly and only official buildings, representative of the French rule, were designed by French architects. Although the regularity characteristic of French urbanism led to design principles which established hierarchies and categories of roads, urban areas and urban functions, architectural eclecticism was the mark of this epoch. Under such vitality, Art Nouveau and Art Déco styles flourished. A simple walk along Carthage, Paris and Liberty Avenues, all perpendicular to the main axis of Bourguiba Avenue, will support this reading. But there are also cases where building and architectural design particularly invited French statements of 'Tunisianity'. Common to the Protectorate period was a deep concern with the spirit of the place. Buildings representing French authority were often conceived as architectural portraits of Tunisia rather than obvious symbols of French dominance. The regional modernism of architects like Raphaël Guy, Victor Valensi and Jacques Marmey is a case in point.[5]

On Bourguiba Avenue, the visitor will first come across the St. Vincent de Paul Cathedral of Tunis, built in 1882 the year after the French took formal control of Tunisia, which incorporates an unusual mix of styles, including Gothic, Byzantine and North African. Nearby, on small perpendicular streets, are the Central Market (1891) and the General Treasury (1900). Then the Municipal Theatre on the main avenue again, an astonishing example of Art Nouveau designed by French architect Resplandy in 1903. Already imposing as such, this theatre used to be part of a bigger complex which included a municipal casino, a minaret tower and a winter garden (*palmarium.*) Long being the gathering *haut lieu* of the European bourgeoisie of Tunis, the complex was eventually destroyed, and only its skeleton has remained behind the 1950s cladding. Bourguiba Avenue is also the ideal place for strolling, especially in spring and summer: the shade of well kept trees, the superb displays of the florists, the dexterity of the shoe shiners, the newspaper stands, light metro stations, the adjacent grand cafes, and the bird song.[6] On both sides of this avenue, the nostalgic stroller would perhaps remember the pre-1956 Résidence Générale de France (now the French Embassy), the Banque de Tunisie, Credit Lyonnais, Compagnie Algérienne, Comptoir National d'Escompte de Paris, the Carlton, Claridge and Tunisia Palace hotels, the Palmarium complex, the cinemas of Le Colisée, Le Capitol, Le Mondial, L'ABC, MidiMinuit,

Le Parnasse, the offices of the famous lawyers, businessmen, insurance and maritime companies, and also the headquarters of the *Dépêche Tunisienne,* the most popular French newspaper of the time which used to brief the public on the latest international and national news on a black board suspended outside its premises.

On Carthage Avenue towards the north, the visitor will discover the Town Hall and the residential buildings of the Rue Ibn Khaldoun, a master piece of Sicilian façading. The Sicilian builders were certainly masters in the art of façade decoration, using lime-based mortar and grey plaster to produce cornices, false capitals, seashell balconies and quasi-floral pilasters (Ambasciata d'Italia Tunisi, 2002). Further up, at the end of Carthage Avenue, across the bridge that spans over the expressway, an old villa which carries a porch with varnished tiles, a local hint, above a potbellied balcony with floral inspirations. Then on his/her way to La Fayette, the stroller may notice the façade of Lycée Bourguiba (formerly the Lycée Carnot[7]) erected in 1931, which calmly presents itself as an Art Déco monument to learning. On La Fayette, the grand synagogue of Liberty Avenue designed by Victor Valensi, inaugurated in 1937 and restored in 1996, is much more imposing. Further up Liberty Avenue is the Belvedere organized as a public park in 1896. Other buildings were on the perimeter: to the south, the municipal slaughterhouse (1888), the tobacco factory and the Italian Hospital; to the west the Technical High School (1906), the Civil Hospital (now Hôpital Charles-Nicolle) and the Civil Prison; to the north the Institut Pasteur (1900). A barren area, between the two city walls to the north of the Casbah was occupied by such administrative buildings as the Directorate of Public Works and Finances (1892), the Directorate of Agriculture and the Justice Palace (1900). The Forgemol Barracks were erected on nearby Bab-Saadoun. These are indeed gems of a beautiful urban design and architecture.

As with many such cities in the Middle East and North Africa, Tunis moved its centre of gravity under Western influence. As a consequence, the economic, professional, and cultural heart shifted to lie between the old medina and the port. Indeed, the principal street of the modern city, the wide, tree-lined Bourguiba Avenue (formerly Jules Ferry, and before that the Promenade de la Marine) appears on maps of modern Tunis much like a battering ram pressing against the vulnerable old city (figure 6.1). This modern part of Tunis holds the world of vehicular traffic, high-rise buildings, a grid pattern of streets and angular buildings – everywhere straight lines and sharp edges, the realm of the geometric, the planned, the practical, and the accessible. Modernization, or more precisely, Westernization, reached Tunis as a cultural area already endowed with a vulnerable, yet established urbanism. The result, especially when it took on the more openly intrusive form of Western colonialism, was a juxtaposition of urban styles which often had clearly defined physical boundaries. With the appearance of the European district in the lower part of the city, and close to the medina, Tunis had a second centre. The continuous development of this modern centre has removed from the medina not only its centrality but its role as a place of exchange.[8]

Suburbia: the Expression of Different Symbolisms

The urban reading of Tunis has to be seen first in relation to the natural obstacles that traditionally hindered a harmonious growth of the city: the Jellaz Cemetery hills and also Belvedere Park and the other zones traditionally unfit for building such as the Sedjoumi Sebkhet in the west, the

Figure 6.1. Tunis showing juxtaposition of old and new city, and Bourguiba Avenue as a battering ram against old city (after Brown, 1973).

Sebkhet Nahli and Ariana in the north and the vast lake of Tunis in the east. After the completion of the European/colonial city, the capital was left with no possible alternatives for urban growth save outside these traditional boundaries – a growth that took the form of planned and non-planned settlements (figure 6.2)

Secondly, in social terms, there are the different forms of Tunis's suburbia. Of these there are the old colonial bourgeoisie quarters as well as villas built on well planned land. The eastern, coastal suburbs (Gammarth, La Marsa, Sidi-Bou-Said and Carthage), with their magnificent landscapes and views, constitute the first element of this category. Here, generally, has lived the old well-to-do aristocracy of the medina. In addition there are the residential quarters of Al-Menzah, Mutuelleville, Al Manar and Nasr. Shaped by the Tunis-Ariana axis and the Airport-Campus axis, this is the site where the emerging upper middle class has lived. Besides the various residential quarters, some of the main monumental projects built here are the Olympic City and the university campus, the Hilton Hotel, the Tawfik Clinic, the Ministry of Foreign Affairs and the still-unfinished headquarters of the Arab League. To the opulence of these suburban realizations is opposed the whole spectrum of urban development undertaken by the government in an attempt to provide functional housing in a country known for its limited resources. The tens of thousands of housing units built in Cité Ibn Khaldoun and Cité Ezzouhour between 1962 and 1976 were meant to absorb the inhabitants of the demolished shantytown of Jebel Lahmar. Then further development followed to absorb the rising lower middle-class population in Ariana, Ras Tabia, Fatouma Bourguiba, Cité Khadra, Manouba, Den-Den, Al-Wardia, Lakania, and Ben Arous.

A further category represents the proliferation of uncontrolled settlements. Although the first peripheral, spontaneous settlements appeared in Tunis in 1930 (in Mellasine, Jebel Lahmar and Saida Manoubia), the most recent ones appeared in the 1970s as a result of both migratory movements from an impoverished countryside, especially from the north-western part of the country after the failure of the 1960s so-called cooperative experience, as well as from residential migrations within the Greater Tunis (Chabbi, 1986). Rough dwellings are built here from cheap materials and are located on small plots of land, generally less than 100 square metres. Between 1975 and 1985, half of the housing units were built in non-planned urban settlements, representing 31 per cent of the total space built, that is 1,600 hectares out of 5,174 hectares. The importance of the peripheral urban space, and its legal status, explains the incredible growth of the informal property market. These lands were traditionally

Figure 6.2. Forms of urbanisation in the Grand Tunis in 1985 (after Abdelkafi, 1989).

known as public *habous* which the state recovered immediately after Independence, and which did not become the object of any registration procedure. Resold to the Medjerda Agricultural Authority, they were then leased to small farmers who slowly came to consider them as their own property.

Spontaneous housing, in its most recent and dynamic forms, constitutes an essential component of urban and suburban spaces of cities in the Arab World. Distributed randomly and informally, it obeys a logic of location and generally develops around areas originally not meant to be urbanized. In Tunis, spontaneous housing settlements encircle the city and constitute a genuine belt which, in moments of crisis, can become a place of social tension. These settlements sometimes consist of small acquisitions spread around clandestine plots, and sometimes form vast concentrations where tens of thousands of people live. This is, for example, the case of Ettadhamen-Douar Hicher in the west of Tunis, which represents the biggest spontaneous settlement in Tunisia. In this area, 500 units were monitored in 1970. Ten years later the number jumped to 8,000 then to 150,000 in the 1980s. An average of 517 ha of land is annually consumed by spontaneous housing; this tendency has been accelerating since the 1980s. What is even more alarming is the fact that a quarter of this is farming land, or at least planned as such in the Tunis *Schéma Directeur*.

The erosion of the city's green belt is happening at two levels: the continuous massive progress of urbanization towards the north and west, from Ariana to Manouba and towards the south in the sector of Mourouj-Ben Arous; and localized, spontaneous urbanization that has a gradual, but harmful effect on the periphery as in Mornag, Khélidia and Oued-Ellil (Chaline, 1990, p. 120). Between 1970 and 1986, the quarters of Mellasine, Saida Manoubia, and Ettadhamen-Douar Hicher included the building of kindergartens, schools, health centres, markets and even sports facilities. The corollary of these improvements is that the municipality has eventually had to take charge of the management of these areas especially at the level of garbage collection.

Lake Tunis: a Third Form of Centrality

Tunis's most ambitious urban and environmental programme thus far remains the *Projet du Lac*. Known as the project of the millennium, it was launched in the 1980s and involved two main operations: the Lac Nord de Tunis and the environmental dredging of Lac Sud.[9]

Since 1988, the Tunisian government has created new institutions and reorganized the institutional framework for environmental protection, the core of which is now the Ministry of the Environment and Land Use Management (created in 1991). This is at the national level. At the level of the City of Tunis itself, the Lake Project (both the northern and the southern parts) has emerged as a logical consequence of the regional ecological situation and its institutional implications at the national level. But there is another factor that should to be considered: the effect of globalization on the urban government of the city of Tunis.[10] One of the key characteristics of the current phase in the long history of the world economy is the ascendance of information technologies, the associated increase in the mobility and liquidity of capital, and the resulting decline in national states' control of key sectors of their economies. Experts and policy-makers appear to be convinced that these aspects have heralded the end of the economic importance of cities. The dispersal capacities emerging with globalization, information technology, the off-shoring of factories, expansion of global networks of affiliates and subsidiaries, the move of back offices to suburbs and out of central cities, led many to suggest that urban economies would become obsolete. Indeed, many of the once great industrial centres, especially in the highly developed countries, did suffer some decline. But, against all predictions, a significant number of major cities also saw their economic power rise, even in the developing world. This is the case of Tunis.

Rather than becoming obsolete due to the dispersal made possible by new technologies, Tunis has become a leading site of finance and specialized services. Although foreign investment in the country started in 1972, the real significance of globalization materialized through the economic association agreements Tunisia signed with the European market. In July 1995, Tunisia entered into an 'Association Agreement' with the European Union (EU) which removes tariff and other trade barriers on most goods by 2008. In conjunction with the Association Agreement, the EU is assisting the Tunisian government's *mise à niveau* (upgrading) programme to enhance the productivity of Tunisian businesses and prepare for competition in the global marketplace. The agreement came into force on 1 March 1998, and its application is judged, overall, to be satisfactory.[11] This new economic role translates into an absolute economic supremacy of Tunis since 90 per cent of jobs in the significant sectors of finance, commerce, economic management, import, and retail are in the capital.

Lac Nord

The trend towards the spatial dispersal of economic and tertiary activities at the metropolitan level, which we have so far associated with the loss of centrality in the medina/European area, has contributed to a demand for new urban development where the needs of the national market could be met. This, obviously, requires considerable physical infrastructure and, together with a growing demand for residential alternatives, is what made the Lake Nord Project not only possible but imperative.

Lake Tunis belongs to a Mediterranean lagoon complex including the Lac Nord, the Lac Sud and the navigation canal. In medieval times, such a geographical setting constituted a defensive advantage. Then it became a barrier to urban development and the growth of the rapidly expanding city. Lake Tunis itself is located east of the City of Tunis and has an area of 37 km^2 and a depth of between 0.5 and 1.0 m. The lake is divided into a north and a south lake by a navigational channel, railroad and a motorway between the city and its second port, known as La Goulette (Halq-al-wady), which is located on the other side of the lake, that is on the Mediterranean Sea. The Lake is a shallow lagoon which was formerly an inlet of the nearby Bay of Tunis. In the course of time, changes in the coastal morphology occurred and a spit was formed across the entrance to the inlet. The formation and growth of the spit continued until approximately the sixteenth century, when the inlet was finally sealed off from the sea. Consequently the water in the major part of the lake became stagnant and polluted, due to the dumping of waste water and sewage. In 1885 a navigation channel was constructed dividing the lagoon into the present Lac Nord and Lac Sud. This channel connects the City of Tunis with the Bay of Tunis at La Goulette. Over the years, the total surface area and depth of the two lakes had gradually been reduced as a result of silting up and of land reclamation schemes.

Direct exchange between the Lac Nord and the Mediterranean Sea was limited to a small canal, constructed in 1951, enlarged in 1978 and dredged in 1981: the Kheireddine Canal. Three smaller canals with culverts connect Lac Nord with the navigation channel. Originally all, but later two of these canals, and the Kheireddine Canal had fishing weirs in them which restricted the exchange of water to some extent. Also, since 1965 the STEG thermal power plant near La Goulette has been pumping its cooling water taken from the navigation channel into the lake. The larger part of the inflow and outflow of the lake water was induced by the tidal movement which shows an average tidal range in the Gulf of 30 cm. For a long time large quantities of raw and secondarily treated sewage effluent from Tunis and surrounding communities, coupled with contamination from urban solid waste tipping areas, poor flushing and circulation, supported dense populations of phytoplankton, abundant growth of *Ulvae* (macro-algae) and *Mercierella* reefs. All this resulted in an offensive smell. To alter this situation, enclosure dams for the western part (where most sludge was deposited) were constructed in 1969, connecting Chekli Island with the northern and southern banks, and isolating the most heavily polluted part of the lake. Later, in 1981, a peripheral drainage canal (ONAS Canal) was constructed along the west bank to receive storm water from Tunis and the effluent and sludge of the Cherguia purification plant and to divert these to Tunis Marine Port by means of a pumping station. From Tunis Port the stormwater flow and the sludge effluent were discharged towards the Bay of Tunis through the navigation channel. During and after long and heavy rainfall the mixture of domestic waste, rainwater and sludge effluent spilled from the ONAS Canal into the lake via three spillways. Furthermore, to improve the circulation of the water in the lake, tidal gates were installed (1972–1974) in the Kheireddine Canal allowing water to flow only into the lake, and in two of the breaches connecting the lake with the navigation channel water was allowed to flow only out of the lake. Despite the removal of most direct sources of pollution, the Lac Nord environment, as a consequence of the build up of nutrient stocks and various remaining sources of pollution, was still in a very bad condition (van Berk and Oostinga, 1992, pp. 23–32).

To remedy these problems, with the help of Middle Eastern investment (Al Baraka Investment and Development) and Dutch Technology, two specialized authorities were set up. The SPLT (*Société de Promotion du Lac de Tunis*) was established in 1983 to manage Lac Nord and, in 1990, the SEPTS (*Société d'Etudes et de Promotion de Tunis Sud*) was created to supervise Lac Sud. The Lac Nord drainage scheme and the development of its banks constitutes one of most ambitious town planning projects of Tunis (figure 6.3). Its goals are: (1) the removal of the contamination of Lac Nord obtaining the contractually agreed upon water quality; (2) the construction of facilities to maintain and guarantee the water quality requirements; (3) the construction of land reclamation and silt deposit areas for, respectively, city/town extension and recreational programmes. Due to an extensive exchange of knowledge, ideas and data between SPLT, the Dutch engineers, and designers and specialists conceptual design was realized within the seven months allowed during the pre-tender phase. Following additional surveys and studies, construction and dredging works started in February 1985 according to the specified time schedule. The 10-kilometre long separation dam was completed in February 1988. The inlet/outlet gates at Kheireddine Canal were completed a month later. Within three years more than 20 million m^3 were dredged, transported, dumped and/or used for 800 ha of land reclamation. Shore protection works with a total length of 35 km were realized. The total cost of the project amounted to 53.8 million US dollars.

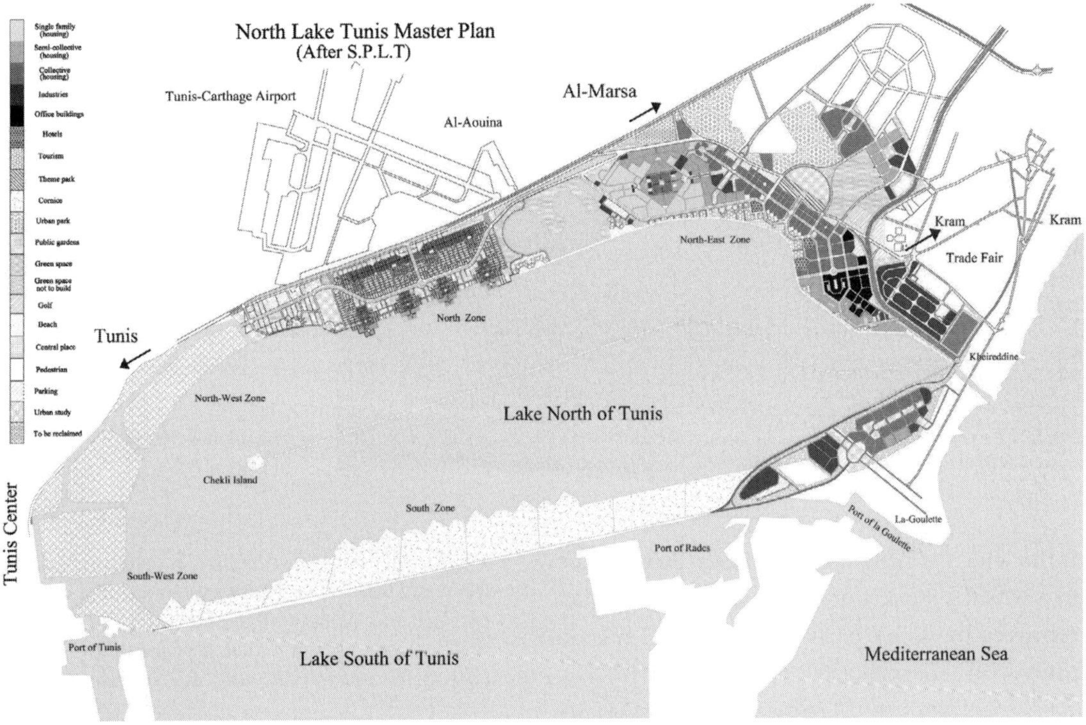

Figure 6.3. Master Plan of North Lake of Tunis (after the SPLT plan).

The first phase of the project, completed between 1985 and 1988, involved the drainage works of the lake and the regulation of the layout of its banks; this was made possible thanks to the state's efforts to restructure the sewage network of Tunis and ensure the interception of polluted waters via the channel built by the National Sanitation Authority (*Office National d'Assainissement*: ONAS). These works put an end to the discharge of waste water into the lake and provided a certain equilibrium to the ecosystem and consequently the possibility to harmonize the northern part of the water stretch in Lac Nord with the surrounding urban fabric. Thus the reclaimed area was integrated with the urban dynamics of Greater Tunis.

As part of the restoration of Lac Nord, SPLT prepared the area surrounding the lake for urban development. Six main areas were identified: the north-west, north, north-east, south-east, south, and south-west banks. The northern zone was developed in the 1990s. Most of this area, known as Al-Khaleej, has been built and inhabited. It comprises leisure facilities, office buildings, residential blocks and single-family housing, and commercial development. This new area has been a success in the eyes of the public (figure 6.4). The Lac Palace and the Dah-Dah theme park, to name just two of these facilities, have become popular landmarks. A new city is taking shape around the shores of a long-forgotten lake, confidently looking towards the Mediterranean Sea, somehow turning its back to the old medina and European city.

Another zone is now being developed. This comprises three allotments on the north-east bank

Figure 6.4. Café on the shores of Lac Nord.

of the lake, covering approximately 205 ha. The area was the subject of a detailed land-use plan approved by decree of the Governor of Tunis on 2 July 1998 and construction started in 1999. The main goals here are: (1) reduction of the pressure on Tunis (hyper concentration of the tertiary services, trade and offices) by the creation of a centre sufficiently well structured so as to provide an alternative to the city itself; (2) structuring of the northern coastline threatened by undifferentiated urbanization, resulting from excessive and rapid building; (3) development of the banks of the lake as a corniche (figure 6.5) by ensuring the continuity of the promenade between the various components of the project; (4) improvement of the natural environment through the transformation of the existing forest into an urban park and the creation of district parks and public spaces. The urban project for the north-eastern zone is neither a housing estate nor a grand urban development but rather a new town type of development, an original planning concept in Tunisia. It aims to ensure the coherence of zoning regulations and the renewal of the urban image of the capital in the second decade of the new millennium. The project includes:

- a polyfunctional town centre associating housing, trade and offices in a typology of collective buildings of 6 to 8 and 10 storey flats with a leisure area along the Corniche, which makes it possible to accommodate, in its central part, hotel and tourist activities;
- a central park and a theme park related to the town centre;
- two districts of collective housing in the east and west of the centre of the town;
- a business and service district related to the neighbouring Al-Kram Trade Center, and a area of non-polluting activities;
- a garden city along the lake front including single-family villas and townhouses;
- a golf course;
- a zone of extension of La Goulette, including townhouses and condominiums;
- public utilities (education, health centres, cultural centres; sports grounds etc.).

The north-eastern zone comprises: Lake Gardens, the non-polluting zone of activities of 'Kheireddine', and La Cité des Pins.

The first allotment is Lake Gardens (106 hectares), located between the banks of the lake on the one hand and the golf course and La Cité des Pins on the other, is primarily for individual housing (figure 6.6). The 428 plots of land, average

Figure 6.5. View from the Corniche.

area 669 m², give to this allotment the aspect of a garden city. It is bordered on the north by semi-collective housing, and on the south by the Corniche and the pedestrian precinct on the lake front, the Corniche being assigned to leisure activities. The heart of the garden city is formed by a quadrilateral of semi-collective and collective housing which also includes public and private facilities along with trade and services.[12] The allotment is divided into seven sectors: Sector 1 (residences of the north); Sector 2 (villas of the north-west); Sector 3 (villas of the north-east); Sector 4 (villas of south-east: lake front); Sector 5 (villas of south-west: Lake front); Sector 6 (the centre); Sector 7 (the Corniche), and functions at two levels, that of a garden city and that of a district. The zones of development are delimited by the roadway system, which indicates the hierarchy of the streets and distribution of the traffic and comprises:

1. A loop made up by Corniche Avenue and the ring road which allows the distribution of the vehicles in all the parts of the residential area.
2. A quadrilateral formed by 18 m wide roadways which allows the distribution of vehicles in the central part of the allotment.
3. A network of 12 m wide roadways which gives access to the residences.

The quadrilateral core is organized around a central square from which one can easily reach the mosque, the administrative services, post office, health centre, and youth club on foot. The Corniche is located some 400 metres away; four car parks were designed to avoid traffic congestion. The purpose underlying the distribution of the different types of housing estates and urban functions is to create a self-functioning district. To reach this objective, allocation of plots has been carefully worked out in order to include: two elementary schools and a secondary college, both located on the ring road which ensures maximum accessibility; two kindergartens and day nurseries, situated on the periphery of the central quadrilateral and connected to the four parks of the districts and to the pedestrian promenades; a mosque, accessible by foot and by car (two car parks being planned for this purpose); the Corniche and parking for 550 cars at the entry of the zone.

The second allotment on the north-east bank of Lac Nord is the activity area of Kheireddine,

Figure 6.6. Lake Gardens (after the SPLT plan).

located within the territory of La Goulette. It covers 44 hectares and is an extension of an existing industrial zone (figure 6.7) This industrial area is near a city park with a direct view onto a stretch of water. It is intended for non-polluting activities (light industry, handicrafts and so on). Shops, restaurants, banks, police stations, a post office have been included so as to ensure social and recreational facilities. Its proximity to the Tunis Carthage Airport and the ports of La Goulette and Radès give this zone excellent communication opportunities.

The third allotment of this north-eastern zone is La Cité des Pins (55 hectares), situated on the municipal perimeter of La Goulette. It is part of the master plan of the north-east zone, alongside the garden city and the Kheireddine activity zone the subject of a detailed land-use plan approved by the Governor of Tunis's decree of 2 July 1998. It covers an area of 552,497 m². The allotment is bounded by the future urban park and the lake shore on the one hand, and Lake Gardens and the future city centre on the other. The allotment essentially consists of collective (condominiums) and semi-collective (townhouses) housing, office and commercial spaces. There are 44 collective housing lots with an average area of 2,718 m², and 23 semi-collective housing lots with an average area of 1,327 m². La Cité des Pins is an urban neighbourhood that will be developed along the axis of the future city centre. Each building will consist of seven or eight storeys. Since this allotment is adjacent to the Lake Gardens sector, the link between the two will be secured through

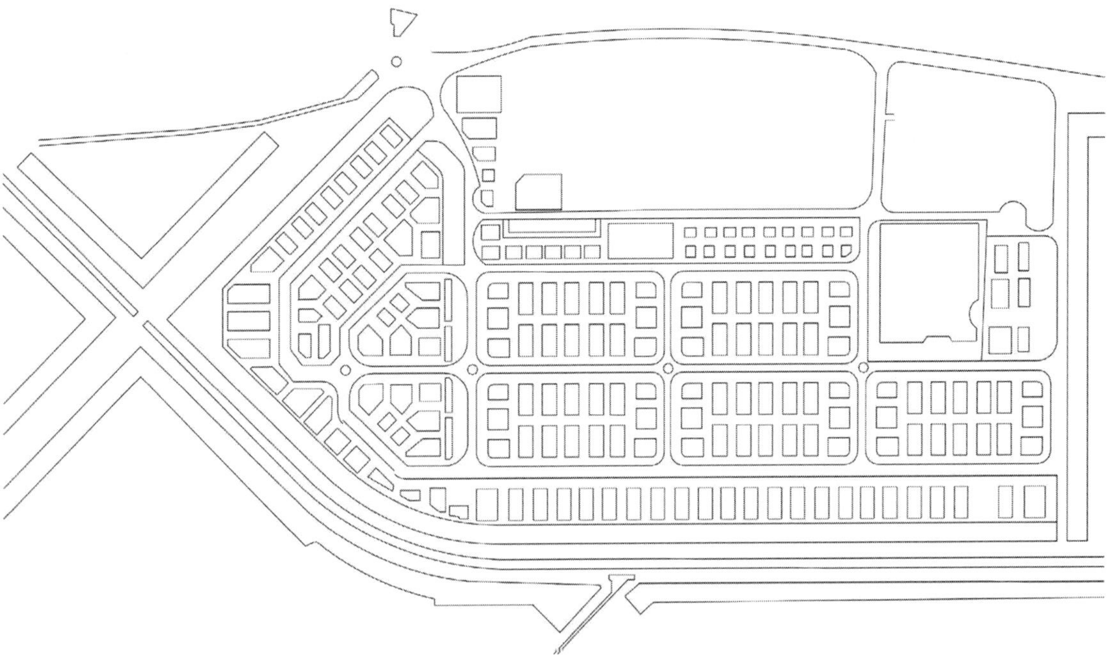

Figure 6.7. Activity area of Kheireddine (after the SPLT Plan).

semi-collective buildings and buildings designed for public use, which are generally four floors high including ground level. On the Corniche side of the allotment a special plot is planned for the building of a high class hotel, itself surrounded by lots destined for tourist recreational activities. La Cité des Pins will not be a dormitory suburb, but will acquire its own sense of centrality. Other zones of Lac Nord, it is hoped, will be developed in the coming years.

Lac Sud

Until the final decades of the nineteenth century, Lac Sud was part of a more extended lagoon with a surface area of 4000 hectares. It was only in 1885, four years after the French colonized Tunisia, that this lagoon was divided into two by the navigation canal which was dredged to give shipping access to Tunis harbour via La Goulette. The south part, about 1,500 hectares with an average depth of 0.80 m, constitutes Lac Sud (figure 6.8).

Given their location in the heart of Tunis, the two lagoons were the city's only outlet for wastewater and therefore their condition deteriorated to such an extent that in 1980 the Tunisian government was forced to initiate a major clean-up programme. This consisted of treating domestic wastewater (previously discharged into Lac Nord) in treatment plants and, between 1985 and 1988, of initiating the Lac Nord development project.

Lac Sud, however, remained in its polluted state until the end of the 1990s. Its location at the centre of the urban and industrial area of south Tunis meant that its ecosystem was constantly under threat. The shoreline of Lac Sud is delimited by a canal destined for the drainage of

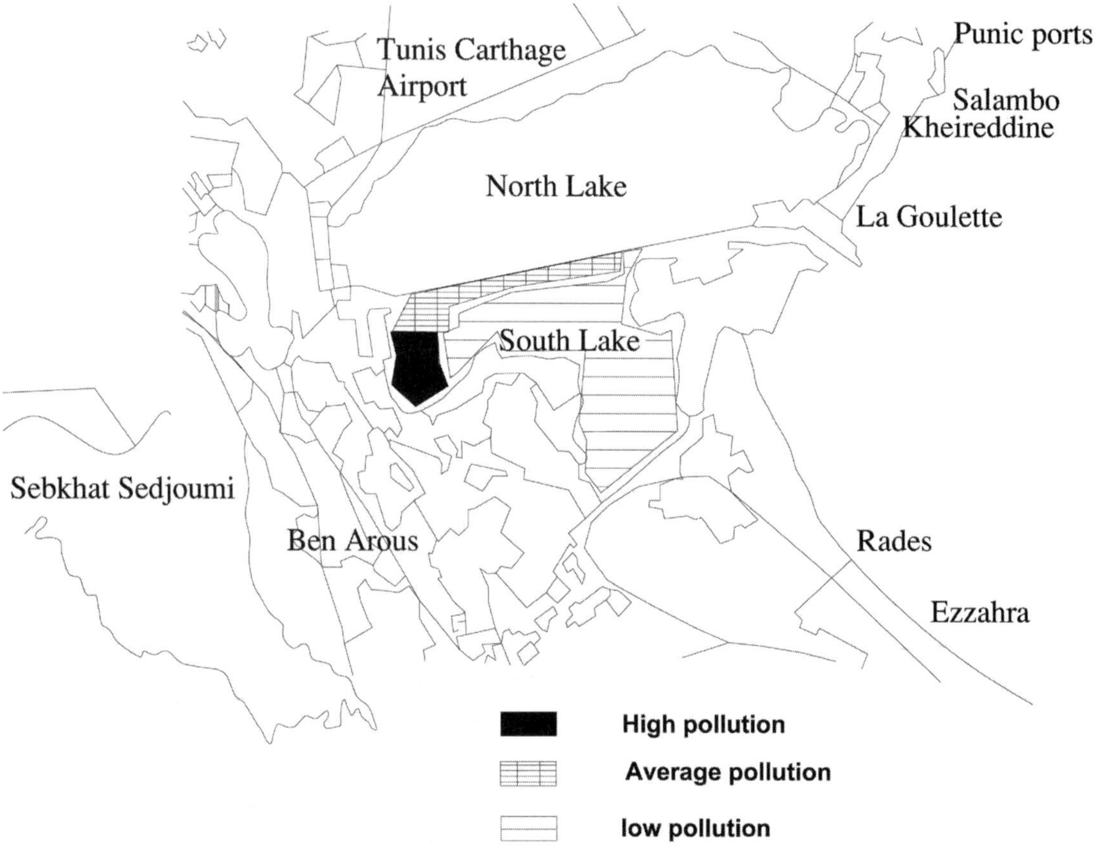

Figure 6.8. Distribution of the pollution within the sediments and localisation of the effluent points in the lake just before the project (after Van Berk and Oostinga, 1992).

rainwater and for the protection of the nearby town of Megrine against flooding. The waters discharged into the lake from the industrial zones of Ben Arous, Megrine, Bir El Kasaa and Radès (a total of about 650 industrial units) have contributed to extreme eutrophication conditions which, in summer, appear with dystrophic crises characterized by red water, bad smells and high mortality of fish life. The rainfall and the industrial waters of these zones are fed into the lake at two different locations: (1) the first at the east section using Bir-Kasaa canal; and (2) the second at the west by Ben Arous and Megrine channels. The industrial flow rates were estimated to be about 3000 m^3/day for the canal of Bir-Kasaa and about 2500 m^3/day for Ben Arous discharge canal. The investigations carried out before the start of the project showed the presence of approximately 2 million m^3 of organic sediment contaminated by heavy metals such as chromium, copper, zinc, iron, nickel, aluminium and also hydrocarbons. A treatment station was set up to treat these waters. During the lake development works, these waters were deviated via a canal towards the Tunis harbour (Vandenbroeck and Ben Charrada, 2001).

In 1990, the Tunisian company *Société d'Etude et de Promotion de Tunis Sud* (SEPTS) was created to promote and develop Lac Sud and its shores.

Since its creation, SEPTS has coordinated different actions with all the parties involved in the zone and has undertaken several studies in an effort to fight pollution in the south zone of Tunis. In 1997, SEPTS offered an international tender for the restoration and the development of South Lake and its shores. In February 1998, the project was awarded to the group LAC SUD 2000 for a period of three years with a lump sum of 60,000,000 euros. The group comprised the following five companies: Dredging International NV (leading company); Van Oord ACZ BV; Tideway BV; Societa Italiana Dragaggi Spa; and Sider Almagia Spa. The Control Inspection mission of these works has been entrusted to a joint venture of consulting firms STUDI (Tunisian) and SOGREAH (French). Theoretically, the remediation of Lake Sud requires reclaiming some 900 hectares of the original 1,500 hectares. The project, known locally as 'the Environment Project of the Century', is certainly one of the most significant environmental projects ever undertaken in the Southern Mediterranean. The new Lake Sud will provide Tunis with some 515 hectares of public green space. The development will create housing (some 23,000 new homes) for the city's anticipated population increase to 2.7 million people by the year 2010. Thanks to this project, Tunis will add 850,000 residents over the next decade, a population increase of some 45 per cent. Lac Sud, which previously covered 1,500 hectares, now extends over an area of 710 hectares and includes a new shoreline of over 13,200 m in length and has an average depth of 2 m. The development works have been implemented within the predicted 3 years.

Conclusion

During colonial rule (1881–1956) the medina of Tunis lost its centrality to the new European zone. After 1956, that is after Independence, the centrality of the modern/European city of Tunis continued, its influence spreading to the adjacent urban areas. But this role of centrality has suffered some setbacks due to increasing problems of accessibility and parking. Some limited attempts at densification have been undertaken with the view of creating a centre alongside Mohamed V Avenue and at the intersection of Bourguiba and Carthage Avenues, where a Kuwaiti firm proposed the building of some tens of thousands of square metres of offices, commercial spaces and hotels by resorting to a more vertical form of urbanism. Despite some limited success, this urban venture, designed by O.C. Cacoub, did not live up to the original ambitions.

The progressive implementation of the light metro services did alleviate the pressure on the centre to some degree, but the problems mentioned above have remained. As a result of this contradictory reality of the modern centre, an increasing process of geographic dispersal has brought about the decentralization of many central functions. The urban areas that have taken most advantage of this decentralization of the tertiary sector are the north and north-east, that is between Parc Belvédère and Lake Tunis. This contributed to the emergence of a third form of centrality in the Greater Tunis basin. With Middle Eastern investment and Dutch technology, two specialized authorities were set up: the SPLT (*Société de Promotion du Lac Nord de Tunis*) to manage Lac Nord and the SEPTS (*Société d'Etudes et de Promotion de Tunis Sud*) to supervise Lac Sud. Despite the apparent success of the Lac Nord project, the significance of its architectural and urban merits is yet to be analysed. In 1995, the critic Serge Santelli highlighted the lack of coherence in the relationship between the road structure and the architectural typologies which led to an unstructured urban landscape that future buildings will not be able to conceal (Santelli, 1995, p. 120). The Lac Sud project is currently being implemented.

The materialization of this new urban reality illustrates the geographic disequilibrium of the city and its history, and also the contradictions between the goals of planning and the weight of market and political decisions in the age of globalization and ecological consciousness.

NOTES

1. The most prominent example is the new US Embassy, designed by Tai Soo Kim Partners, an architectural firm based in Connecticut, and local architect Wassim Ben Mahmoud. Other diplomatic missions that have moved to the Lac area include the Lebanese and Indonesian Embassies.

2. Among the famous visitors of the time: Alexandre Dumas, Gustave Flaubert, Guy de Maupassant and, later, André Gide, Paul Klee, Kandinsky, Maria Rilke, and the Russian painter Roubstoff.

3. The exodus from the old medina may be said to have begun as early as 1865, even before the French Protectorate, when a Westernizing Tunisian minister (Kheireddine Pacha Tounsi) established, in the vicinity of the Casbah, the Sadiqi College as an institution specifically intended to train government officials by means of a combined Western/indigenous curriculum. Kheireddine Pacha was forced into exile in Istanbul in the mid-1870s, just a few years before the French entered the country in 1881.

4. When the Jews were expelled from Spain in 1492, and a few years later from Portugal, they sought refuge where they could. Large numbers went to Morocco, Holland and the welcoming Ottoman Empire. Others ventured further afield, to the Caribbean and the New World and others as far as Eastern Europe. Therefore, when in 1592, Ferdinand I (di Medici), Grand Duke of Tuscany, invited them to settle and freely practice their religion in Pisa and Livorno, many expelled Sephardim eagerly took him up on it. Thus was established the community of Sephardic Livornese Jews. Trading from these Italian ports, they spread out through the Mediterranean and in the eighteenth century some settled in Tunis, as trading agents for their relatives in Livorno. Known as *grana* (plural of *gorni*, that is a *livorno*) they found little in common with the local Jews, *touansa*, and kept apart from them, maintaining separate synagogues, cemeteries, and kosher rules. Even politically they disagreed, the Livornese favouring a Protectorate under Italy while the locals favoured France.

5. See Breitman (1984) and Santelli (1995).

6. Some of the pre-1956 cafés on Bourguiba Avenue, former Jukes Ferry, included: le cafe Casino, le cafe de Paris, la Rotonde, Le Baghdad, Le Capitole, la Patisserie Royale, Chez Max, Le Novelty, Le Paris Bar and others.

7. Lycée Carnot has some history to it. In 1875, some Catholic missionaries in Africa were invited to Tunisia as guards of the Chapel Saint-Louis, which the French government had built in 1830 in Carthage, the place where it is thought Saint Louis died during one of the Crusade wars in the Middle Ages. In 1880, Cardinal Lavigerie erected, in the proximity of the chapel, several monumental buildings where he built a boarding school which he called Collège Saint-Louis de Carthage, and where in 1881 there were more than 50 pupils. The college was then moved to Tunis itself where it took the name of Collège Saint-Charles. This establishment slowly lost its religious character. In order to secure a respectable level of education, the cardinal requested the Ministry of Public Education to provide him with university teachers. Lay professors, paid by the Tunisian government, were accordingly assigned to the task. In 1889 the college was donated by the cardinal to the Tunisian government. It then took the name of Lycée Sadiki, because it was decided that some students from the famous Muslim College Sadiki, built by Kheireddine Pasha in the 1870s, would be transferred to the new college in the European city. It kept this name until it was reorganized by the decree of 29 September 1893. Immediately after the death of President Carnot, it was decided in the Council of Ministers that the college take the name of Lycée Carnot.

8. Admittedly, some of the areas outside the circle enclosing the old medina reveal signs of the old forms and patterns. Such areas might better be labelled transitional or hybridized, since they reveal interesting combinations of Western and indigenous forms, but these are neither so extensive nor so important as the thoroughly Westernized areas. Traditional zones like these are, in any case, satellites of the new city, responding to imperatives and attractions lying outside of the old medina (Mutuelleville, Bardo, Franceville, Montfleury, Bellevue).

9. Other sanitary operations have also taken place in Greater Tunis, namely the Tunis Sewerage and Reuse Project and the Lac Sejoumi Sustainable Project.

10. Some quick facts: the total population of Tunisia is around 10 million. In 2001, the average per capita income was $2500 per year. The total workforce in 2001 was 3,292,700. Services constitute 44.1 per cent of the economy; industry 33.2 per cent; agriculture 22 per cent. Greater Tunis currently has some 2 million inhabitants.

11. Tunisia is thus the most advanced of the Euro-Med partners as far as the introduction of a free trade area with the European Union (with a twelve-year transitional period) is concerned. The country started dismantling tariffs in 1996, that is to say, before the EU-Tunisia Association Agreement came into force in 1998. It applies the agreement satisfactorily overall, but there are occasional problems in certain areas. Tunisia has also concluded free trade agreements with Egypt, Jordan, and Morocco. Tariff dismantling has seen a speeding up of the country's integration into the European market and 80 per cent of Tunisian exports go to the EU, with the EU providing 71 per cent of Tunisia's imports.

12. See Société de Promotion du Lac de Tunis (SPLT) (1998).

REFERENCES

Abdelkafi, J. (1989) *La Médina de Tunis*. Paris: CNRS; Tunis: Alif .

Abdelkafi, J. (1997) Tunis et son lac. *Architecture Méditerranéenne: Tunisie*, pp. 150–157. (No volume number.)

AlSayyad, N. (ed.) (2001) *Hybrid Urbanism: on the identity discourse and the built environment*. Westport, CT: Praeger.

Ambasciata d'Italia, Tunisi *(2002) Architectures italiennes de Tunisie*. Tunis: Finzi Editeur.

Amodei, M. (1984) Tunis 1860–1930: The formation of a colonial town. *Environmental Design: Journal of the Islamic Environmental Design Research Centre*, **1**, pp. 26–35.

Beguin, F. (1983) *Arabisances, limites and grand tracés*. Paris: Dunod.

Blake G. H. and Lawless R.I. (eds.) (1980) *The Changing Middle Eastern City*. London: Croom Helm.

Breitman, M. (1995) Rationalisme et tradition : Tunisie 1943–1947. Paris: Institut français d'architecture.

Breitman, M. (1984) J. Marmey: between rationalism and tradition, tradition and Moroccan works 1938–1955. *Environmental Design: Journal of the Islamic Environmental Design Research Centre*, **1**, pp. 6–13.

Brown, Carl, L. (ed.) (1973) *From Madina to Metropolis: Heritage Change in the Near Eastern City*. Princeton, NJ: The Darwin Press, pp. 28–30.

Chabbi, M. (1986) *Une nouvelle forme d'urbanisation à Tunis, l'habitat spontané péri-urbain*. Paris: IUP.

Chaline C. (1990) *Les villes du monde arabe*. Paris: Masson.

De Koninck, H. (1991) Les villes du patrimoine mondial: capitales du temps. *Cahiers de Géographie du Québec*, **35** (94), pp. 9–87.

Duchac R. (1974) *Villes et sociétés au Maghreb : études sur l'urbanisation*. Paris: CNRS.

Hakim, B.S. (1986) *Arabic-Islamic Cities*. London: KPI, pp. 56–60.

Jbili, A. and Klaus, E. (1996) The Association Agreement between Tunisia and the European Union. *Finance & Development* (Quarterly publication of the International Monetary Fund and the World Bank.), **33** (3), pp. 18–20.

Kassam, A. and Zghal, A. (1980) *Systèmes urbains et développement au Maghreb*. Tunis: CERES Production.

Revue Architecture Méditerranéenne, numéro special, Tunisie (1997). Marseille. (No volume number.)

Santelli, S. (1995) *Tunis, le creuset méditerranéen*. Paris: Editions du Demi-cercle/CNRS.

Société de Promotion du Lac de Tunis (SPLT) (1991) *Lotissements A et B de la zone nord des berges du Lac de Tunis. Lotissement El Khalij*. Tunis: SPLT.

Société de Promotion du Lac de Tunis (SPLT) (1998) *Plan d'Aménagement Général*. Tunis: SPLT.

Van Berk A. H. and Oostinga H. (1992) North Lake of Tunis and its shores: restoration and development. *Terra et Aqua, International Journal on Public Works, Ports and Warer ways Developments*, No. 49.

Vandenbroeck, J and Ben Charrada, R. (2001) Restoration and development project of South Lake of Tunis and its shores. *Terra et Aqua*, No. 85.

Chapter 7

Cairo's Urban *Déjà Vu*: Globalization and Urban Fantasies

Khaled Adham

The effects of recent privatization and general 'globalization' of capital and society in Egypt as well as the ensuing urban and real estate explosion of Cairo, particularly of new exclusive urban spaces of consumption for the elite segment of society, have a sense of newness about them.[1] Close observers of Egypt's modern history, however, experience a *déjà vu*: has Cairo been down this road before, or is this an untrammelled route in the city's socio-economic and urban narratives? There are indeed striking similarities between the trajectory Cairo has traversed at the turn of both the twentieth and the twenty-first centuries. At the turn of the twentieth century, following the *dénouement* of the days of Egypt's receivership, brought about by Khedive Ismail's debts to foreign European powers, the Egyptian government launched rigorous privatization and economic liberalization initiatives. These undertakings culminated in the sell-off of the largest state-owned company, *Daira Sanieha*, to foreign financiers and the launch of the National Bank of Egypt (Raafat, 1997). Following this shift to market capitalism, the government proceeded to improve the urban infrastructure, particularly, of the capital city. These new developments sparked a seemingly unprecedented era of economic prosperity not experienced in Egypt's modern history (Goldschmidt, 1988). Capital flow and economic integration with European markets were followed by an influx of foreign businessmen, speculators, real estate developers, and architects. In the first decade of the twentieth century, not only European-style restaurants and casinos proliferated in Cairo, but also new, exclusive real estate developments, such as Maadi and Heliopolis. The former emulated the European environment and lifestyle whereas the latter reflected the growing European fantasy of the Orient during the colonial era. As diverse as they were, these developments represented a new urban economy mirroring the consumption patterns of the dominant, growing sectors of agrarian capitalists, white-collar government employees, and foreigners, who had begun to establish themselves in new urban areas inside and around Cairo. It is important to note that the economic boom was short lived, however, and by 1907 a financial crash, caused by over-speculation in real estate, slowed down the building construction fever.

Fast forward almost a century and many of the circumstances that ushered the economic upsurge of the 1890s resemble those that led to the upswing of the 1990s.[2] For example, Egypt's fiscal turnaround started with the debt forgiveness which resulted from a political decision of the United States and its allies during the Gulf War, and was followed by an aggressive privatization programme. As in the beginning of the twentieth century, the recent economic boom was accompanied by the sprouting of international fast-food chains, theme parks, golf courses, shopping malls, and the most dramatic real estate explosion Cairo has ever witnessed.[3] Akin to the earlier boom, the latest involved improving and developing the infrastructure in and around the city to facilitate the real estate expansion. In a matter of a few years, the vast deserts surrounding metropolitan Cairo became dotted with satellite cities and lavish, exclusive, gated communities, such as Beverly Hills, Rehab City, and Dreamland. All these urban spaces are connected with highways and a ring road circumventing the crowded, sprawling megalopolis. In what may seem, at least on the surface, a reverse order to Heliopolis, many of these developments reflect the dominant social group's increasing infatuation and fantasy with Western, particularly American, ways of life, an urban Occidentalism, if you will. In an interesting continuation of this parallelism with the turn of the twentieth century, we are currently experiencing a drastic slow down in the construction businesses caused by over-speculation in the real estate market.

While granting the apparent similarities of these two epochs, drawing parallels must still be undertaken cautiously, for the possibilities and range of positions and interpretations that lie in this direction are numerous.[4] My concern is that in many comparative studies of the built environment, any semblance between two different eras may lead to a glib equation, or even conflation of many contemporary concepts and discourses with older ones. For instance, if 'globalization' is used nowadays by some to mean the effects brought about by transnational capital and culture on local economy and society, then it is possible to question whether Cairo experienced globalization at the beginning of the twentieth century. The problem arises not when a contemporary concept such as globalization is taken to examine an earlier epoch, but when it becomes subsumed by other historicized concepts such as modernization, Westernization, or any other term used to depict the changes taking place a century ago. Conflating words confuses and evades meaning, particularly because most of these words and terms are not innocent or neutral in their usage, and are always socially as well as politically constructed to evade, ameliorate, obfuscate, or replace other discourses, such as exploitation, domination, or imperialism. My point is that these terms are very contextual, hence, contingent on the political and economic situations of the time and place in which they were used. For many people around the world globalization is increasingly becoming synonymous with Americanization, American hegemony, or neo-colonization by the Anglo-American axis.

Collating the economic, political, social, and urban issues of Cairo across the time span of the twentieth century is, therefore, intertwined with some present day academic and popular cultural debates and conflicts, in which the concept of globalization plays an indirect, but very important role in the discussions. It should be clear from the outset, however, that my goal in this chapter is not to engage in debates concerning the merits of globalization or liberal economies – even though it may be unavoidable to shift away my own prejudices as an inhabitant subject of the contemporary Middle East. Nor is my goal to deploy a survey of the urban economics of the city in these two epochs – this effort goes beyond the scope of a single essay. Rather, my main effort is to provide through a back and forth excursion in historical space and time, and through the delimiting view of my contemporary eyes, an overall view of the influence that international capitalism has exerted upon Cairo in the opening and closing decades of the twentieth century. Specifically, I shall discuss the interrelationship between landed capitalism and the emergence of a new urban economy based on new patterns of consumption, leisure, and fantasy in Cairo – a new lifestyle. To do so, I shall outline how the incorporation of the Egyptian culture and economy into the world capitalist, socio-cultural and economic systems influenced the production of two selected land developments, urban fantasies, from these two eras, namely, Heliopolis and Dreamland. While the former is one of landed capitalism's earliest fruits in Cairo, the latter is one of its latest fads. In the context of this study, it is important to denote that the seeds of landed capitalism in Cairo were sown during the emergence of Egyptian capitalism in the nineteenth century. Whereas in this short history of free market in land, the first decade of the twentieth century represented the relationship between international capitalism and the production of spaces during Cairo's colonial past, the last decade signifies this dialectic at work in the city's postcolonial present and future. Both epochs, therefore, reflect significant moments, ruptures, or histories of the continuous and complex inter-actions between Egyptian and world socio-cultural and economic systems as they have been played out in the production and commodification of spaces in the capital city.

A critical reader may, rightly, question the degree to which these two land developments – Heliopolis and Dreamland – accurately represent the urban and spatial conditions in Cairo for the first and last decades of the twentieth century, respectively. I have to confess that any claim to representation involves repression: some things are said and brought to fore, others are silenced and repressed (Chambers, 1990). My narration of these cases will therefore admittedly remain limited by the constraints inherent in any act of representation. In addition, few scholars today would repudiate that the patterns and dynamics of any city's spatial expansions are diverse and produce a series of disconnected, heterogeneous geographies and not an unvariegate space. That is why despite the similarities we may find within these two epochs, any detailed examination of selected cases will remain limited and will not be wholly representative of the urban and spatial conditions of Cairo. Heliopolis and Dreamland are unique, concrete human experiences from two different eras and my selection of them depends upon three similarities. First, both cases are large real estate developments produced by the private sector for money profit. Second, they are, arguably, presented by their makers as utopian visions, urban fantasies, for a total environment that would resolve the urban ills and problems of Cairo in their respected times. Third, both developers selected the surrounding desert of the metropolitan area for their projects. This being said, my discussions of these two epochs will undoubtedly transgress the boundaries of the two specific cases to other 'urban fantasies' from their respective times.

Methodologically, the inquiry into the interrelationship between international capitalism and urban fantasies in Cairo during these two epochs is based on three premises. First, I take the semiotics of space and urban fantasies in Cairo as representational dimensions of class, lifestyle, consumerism and other forces in the history of

Egyptian capitalism. Second, a variety of contemporary theorists have suggested that contemporary 'global' culture is in a lingering threshold between two eras: modernity and postmodernity (Woods, 1998; Kellner, 1997). This condition of in-betweenness means that the relationship between capitalism and the city can be articulated with theories from either side of the academic debate, modernity or postmodernity.[5] Third, while the observation of this historical semblance regarding the first and last decades of the twentieth century rests on a key theoretical association between, on the one hand, the economic activities and dominant mode of production and, on the other hand, urbanization and the spatial ordering process of the city, I do not necessarily take its default models for this theoretical association at their face value. I agree with Dona Stewart that most urban theories and frameworks have been proposed with a default model, the Western city, in mind, but non-Western cities 'engage the world capitalist system in dissimilar manner' (Stewart, 1999, pp. 128–129) To a large extent, this default model relies on some recognizable, if not formulaic, character. Sanjoy Chakravorty makes similar remarks in her study of Calcutta (Chakravorty, 2000). For example, in examining non-Western contexts, she argues that scholars always take the two-stage distinction between Fordist and Post-Fordist economies as the crucial element separating past pre-industrial from present industrial modern economic activities. Balancing the economic variable with the political, I identify five distinct stages for the study of the relationship between international capitalism and Cairo. These stages correspond to the historical stages of capitalist development as a world system as well as political and economic changes specific to Egypt: (1) colonial penetration (1798–1882); (2) colonial domination (1882–1952); (3) postcolonial nationalism (1952–1974); (4) postcolonial penetration (1974–1989); and, (5) postcolonial domination (post-1990). The observation concerning the sudden expansions of Cairo corresponds with the second and the last stages. This does not mean that Cairo did not expand significantly before or between these epochs.[6] What makes these two identified epochs or stages in the twentieth century different from other earlier expansions, however, is the relationship between their socio-political and economic orientations and the globalizing, expansionist development of Western capitalism in the last two centuries.

At this point, I would like to explain how I use the word globalization, rather than continue to bandy it about, knowing full well that the word could have as many meanings as users. How I use globalization will also constitute the method I shall deploy to collate these two eras and land developments within them. I shall follow Arjun Appadurai and argue that the current world culture must be seen as a complex overlapping, disjunctive order (Appadurai, 1994). Appadurai proposes a framework comprising five dimensions of intercultural flows that would make the order of 'globalization' more understandable and convenient to collate across time, namely, ethnoscapes, technoscapes, financescapes, mediascapes, and ideoscapes (Appadurai, 1994, pp. 324–339). Paul Knox presents a concise summary of these ideas and adds to them a sixth dimension:

> 'ethnoscapes' (produced by flows of business personnel, guestworkers, tourists, immigrants, refugees, etc.); 'technoscapes' (flows of machinery, technology, and software produced by transnational corporations and government agencies); 'financescapes' (flows of capital, currency, and securities); 'mediascapes' (flows of images and information through print media, television, and film); and 'ideoscapes' (flows of ideological constructs, mostly derived from Western world-views: e.g. democracy, sovereignty, representation, welfare rights, etc . . . 'commodityscapes,' produced by the flows of material culture that encompass everything from architecture and interior design through to cloth and jewellery. (Knox, 1995, p. 6)

It is important to denote that by using the suffix 'scape' Appadurai wants first to emphasize the fluidity and irregularity of these constructs and, second, to stress their subjectivity. I shall take the discussion of these dimensions as operational variables or vehicles of 'globalization' to move back and forth across the span of the twentieth century as well as across the metropolitan city from Heliopolis in its north-east to Dreamland in its west.

The Colonial Economy and Agrarian Capitalism

The process by which capitalism emerged in Egypt is a complex story that can only be sketchily dealt with here. In part, the process must be recognized as having roots that go back to some sporadic events during the nineteenth century. Before this time, Egyptian society lacked some of the common features that define what economists and historians would describe as capitalist society. True, there might have been some capitalist transformation of the Egyptian economy by Muslim merchants and Mamluk rulers during the eighteenth century, as Peter Gran has argued convincingly (Gran, 1997). There is no doubt, however, that some prerequisites for a modern capitalist system began to appear and accelerate through the integration with Western Europe in the nineteenth century. These prerequisites revolve around the presence of the modern institution of private property and market system with its peculiar relations and factors of production: land, capital, and labour.

The history of private property in Egypt is vague and confusing to say the least. Timothy Mitchell contends that there is a silent history over the conditions that produced capitalism in Egypt, and argues that this might have been a necessity for power to operate and expand within the system (Mitchell, 2002). Surely, part of this confusion stems from the fact that scholars have applied their contemporary understanding of capitalism and private ownership of land on past forms, hence, framing the meaning with our contemporary biases and views. For example, scholars differ over the meaning of landholding during the nineteenth century: was it an object with a single claim, or was it, as Mitchell tells us, 'a system of multiple claims'? (Mitchell, 2002). No doubt that the 1882 British occupation of Egypt and the subsequent need to tax farmers and landowners was an important milestone in the history of property rights. Of course, at different points in Egyptian history, the right of some individuals to own property and wealth was recognized. But the right of ownership to all persons was never accorded until well into the nineteenth century. The peasants who worked on agricultural land, for instance, rarely owned it. Mohamed Ali, the ruler of Egypt between 1805 and 1849, announced at the beginning of 1814 that all the land of Egypt belonged to the ruler, and hence abolished even the limited ownership of land that existed under the *Iltezam* system of the eighteenth century.[7] Although Ali granted some land to relatives, high-ranking military officers, Bedouins, and to the *mashaikh*, or heads of extended families, that collected the taxes from the peasants, not one of these landholders had the absolute authority over the land.[8] For instance, they could not sell the land and some were not able to pass it to their siblings. Moreover, under Ali's land tenure system, the central government decided every minute thing concerning the land, from the type of fertilizers used to what the peasants sowed. For Ali's political power as well as his expansionist ideas, land was regarded as the basis for military power and civil administration through a particular system of control and taxation. Land was therefore not yet considered a real estate to be bought and sold, or developed for money profit in the modern sense.

In addition to the recognition and the institution of private property, for a fully-fledged capitalist system to emerge in Egypt, the development of a modern market system was required. Of course, there had always been a market place in medieval Cairo where merchants sold their goods. But the market system where land, capital, and labour were considered commodities to be exchanged for some abstract value began to develop only later, at the end of Ali's reign and during the time of the subsequent rulers.[9] One significant event that marked these loose beginnings of the shift took place with the reform land laws in 1858, known as the *Sai'idiya* Code. The Code 'recognized hereditary rights to agricultural land formerly considered part of the state domains, thus facilitating the extension of private freehold ownership over much of the country' (Abu-Lughod, 1971, p. 152). In addition, under the Code foreign nationals were given the right to purchase and own land.

This being said, signs of the road to a capitalist economy may have begun to manifest even a decade earlier in the financial sector, particularly in 1843, when the government approved the establishment of the Alexandria Bank. We know that as early as 1837, Ali recognized the importance of banks and modern financial institutions as the steadiest sources of capital to develop the country. It is worthwhile noting that before the advent of institutional banking, usurious moneylenders, though prohibited by Islamic religion, were traditionally supplying peasants and city dwellers with cash, while royalty and nobles borrowed from wealthy merchants (OBG, 2001, p. 65). The impact of the Alexandria Bank was not so significant, because it was solely involved in discounting treasury bonds issued by the government; it did not have sufficient cash in reserve to carry out other banking operations. And although the bank was not successful, shortly afterwards, in 1855, another bank, led by a foreign consortium and entrepreneurs, was founded. Like the earlier bank, this one too was little more than a joint-stock commercial enterprise with very limited funds, mostly tied to a single loan for a single member of the royal family, and thus, not unlike the earlier bank, it also foundered. The presence of a steadier and a truly banking environment would thus have to wait for another twenty-five years. I shall return to this point later.

During the liberal reign of Khedive Ismail, who took over from Said Pasha and ruled until 1879, Egypt became increasingly tied culturally, politically, and economically with Europe. This was so partly because of the cotton boom that began in 1861, when Egyptian cotton became the premier source for Europe's textile industry, and partly because of new geographic conditions brought about by the Suez Canal project.[10] It might be accurate to describe the opening of the Suez Canal in 1869 as one of the most significant events in Egypt's modern history. Besides channelling a vast flow of international traffic through the country, increasing the linkage with the outside world, the opening of the canal drew Egypt into the geopolitical, colonial, struggle among European powers, thus, closer to the system of expanding Western capital. For Cairo, the event has been hailed as a benchmark for the beginning of the Westernization of the city, and for our concern here, it marks the inception of the process of space commodification in the capitalist sense. The Khedive was determined to make the event a spectacle before the eyes of the many European dignitaries whom he had invited – more spectacular than the, then, fashionable World Expositions. To boost his own image as a ruler of a modern, civilized nation, Ismail wanted the city to be developed and modernized along Western lines and, therefore, Cairo had to be a graft from the new urban design and architectural styles of Paris. Functioning like a vast developer, the government selected a 1 mile by 2 mile tract of land between the river Nile and the medieval city and launched a massive development project that would alter the face of the city forever.

In sharp contrast to the native, pre-industrial city, from which visitors could still see the silhouette of its skyline, the new Ismailia quarter was planned to include open squares that were adorned with statues, gas-lit avenues with sidewalks, underground sanitation, an opera house, gardens, pavilions, and palaces. The site was swiftly laid out and plots were given away to any one who was willing to spend a huge sum of money on a villa and finish it before the opening of the canal (Rodenbeck, 1998). During this time, members from the royal family entered the real estate business, including Khedive Ismail himself, who bought large tracts in the area, subdivided it, and then sold it to others for profit. In fact, the practice was so profitable that the Khedive emulated it in the newly developed suburban and winter royal retreat of Helwan. (It is important to note that these real estate ventures by the royal family continued throughout the rest of the nineteenth century during the reigns of subsequent rulers. Later, for instance, at the end of the century, Khedive Abbas Hilmi II went even a step further and planned, built, and sold the Khedive Apartment Buildings in Emad el-Din Street, in the Tawfiqia quarter, north of the Ismailia quarter (Muselhy, 1988, p. 228).) The gradual emergence of the European-like quarter to the west of the old city paralleled the gradual rise of the Egyptian bourgeoisie and an influx of foreigners. The new quarter accommodated these new urbanites and spatially facilitated the emergence of modern, European, urban lifestyle.

The lavish public and private expenditure during these days sky-rocketed Egypt's debts to foreign creditors and bankers, plunging the country into receivership and foreign control, particularly after the government was forced to declare bankruptcy. The European colonial penetration of Egypt became eventually a political, colonial domination when in 1882 the British military forces occupied the country. From this date, Egypt was ruled, in name, by a Khedive, but in reality, by the British counsul-general. A duel French-British control system of administration was installed to oversee the Egyptian economy, which would allow European creditors to ensure that the debt was paid off.

With a snooty counsul-general running the country as a business and whose simple objective was to pay back the debts to European creditors, the Egyptian economy became increasingly integrated with and subordinated to the expanding transnational European capitalist system. To Egyptians, the economic integration with Europe meant that in addition to the export-oriented economy and foreign capital flow in the country, there was an influx of foreign people, goods, ideas, and technologies. There is a great deal involved in this story of integration and subordination – of the changing ethnoscape, commodityscape, ideoscape, and technoscape – and, of course, many ways to narrate it, but I select one strand from this complex tale.

Timothy Mitchell tells us that the establishment of the department store in Europe played a crucial role in this drama (Mitchell, 1991, p. 16). The department store – a highly concentrated place for consumption which first appeared in Paris in 1852, Bon Marché, and then mushroomed in Europe and the United States – introduced the ideas of stockpiling, advertising, and fashion to the increasingly eager-to-consume urban bourgeoisie (Pevsner, 1976, p. 267). (It is important to note that the department store was a development of earlier building types, the arcade and the galleries, like the Royal Opera Arcade built in London in 1818 and the Galerie d'Orléans built in Paris in 1830, respectively. The arcade, so slightly different than the closed Parisian version, would appear later in Cairo, in Mohamed Ali Street, and eventually in Heliopolis.) The creation of the department store resulted in an unprecedented increase in profit that impetuously boosted the textile industry, which, in turn, demanded a further increase in the cultivation of cotton. Egypt's task became then to focus solely on producing more raw cotton for the British textile factories, something that the counsul-general, Lord Cromer, had guaranteed already. According to Beinin and Lockman, the rapid expansion of cotton and its export to Europe 'provided much of the impetus for the transformation of agricultural land into private property, a transformation that resulted in the restructuring of agrarian social relations' (Beinin and Lockman, 1998, p. 8). An export-oriented, single crop, economy, therefore, gradually replaced the subsistence economy.[11]

Gradually also came the removal of all state restrictions on private ownership of land for both nationals and foreigners and the authority to mortgage land, to pledge it legally against a loan. At this moment, land was treated as a commodity to be exchanged in the market, or as a source of income and profit (Chichian, 1988, p. 28). In fact, around 1894, one-third of the agricultural land was held as private property and the number of landowners climbed to a record high. Parallel to these changes in laws were gradual improvements in the system of recording land titles. Abu-Lughod tells us that during the 1880s 'a totally revised system of land ownership, in which the unchallenged dominance of private property provided a marked contrast to the state monopoly system that had prevailed under Muhammad Ali' (Abu-Lughod, 1971, p. 152). Thus, it was not a coincidence to find, at the same time as the improvements in property rights and title recordings, the emergence of mortgage banks that specialized in loans against land security. It was first a French mortgage bank, with French capital, which began operations in 1880; upon its initial success the door was opened wide for other European financial institutions to follow. Thus, banking activities emerged and grew spectacularly for the first time since the 1850s, and Chichian quotes Crouchley remarking that during this time 'eight banks provided telegraph exchange in Paris and London and two foreign mortgage banks were opened in 1880' (Crouchley, 1936, p. 28).

But perhaps the most significant year for the Egyptian economy and the *laissez faire*, liberal economic policy was 1898. In that year two significant, interrelated events occurred: the founding of the National Bank of Egypt and the selling off of the *Daira Sanieh*, the largest state-owned company, to a private consortium (Raafat, 1997, p. 17). Samir Raafat contends that these two events sparked this unprecedented economic well being that, in the words of another historian, was 'not experienced since the Mamluk times' (Goldschmidt, 1988, p. 48). Soon afterwards, particularly in 1903, another important capitalist institution was established: the first Cairo Bourse. In a revealing gesture to the new economic orientation of the country, the site selected for the 'Bourse Khediviale du Caire', was the centre of Cairo's European district, Ismailia. The capitalization of the Cairo and Alexandria stock Exchange soared in a matter of a few years and, by 1907, 'the number of companies trading in the Cairo Bourse alone had reached 228 with a combined capital of L.E. 91 million' (Raafat quoted in Shahine, 1999, p. 14). Along with the expansion of capitalist institutions and rapid growth of the Egyptian economy, a considerable influx of foreigners to Cairo facilitated the flow of foreign capital. Foreign capital investment increased immensely during the first decade of the twentieth century with a high concentration on land speculation and mortgage (Chichian, 1988, p. 33). According to Charles Issawi, there was a spectacular investment boom so that 'between 1900–7, 160 companies with a capital of L.E. 43 million were formed' (Issawi, 1954, p. 39). Moreover, almost all the total value of the joint-stock companies was owned by French, British, or Belgian nationals.

Add to this economic ferment at the turn of the century, the re-conquest of the Sudan, the completion of the Aswan Dam, and the declaration that Egypt was solvent and one gets a picture of the optimism that prevailed among the foreign and national investors alike (Marsot, 1985, p. 78). In the short term, the influx of foreign capital inflated the economy and the stock market, yielding a very high return on investment to the speculators. This, in turn, attracted even more investment in real estate. Abu Lughod, once again, presents a concise summary of the condition during this era.

> Since much of this new capital was in the form of loans on land rather than investment in productive facilities, it was natural that land values should have benefited most dramatically from the heightened competition. While agricultural lands, particularly those enhanced by the new irrigation system created by the Aswan Dam after 1902, experienced some of the sharpest upturns, all forms of speculation in land and buildings – especially on the lands outside of Cairo and Alexandria – yielded substantial profits. Land and building societies were founded in increasing number between 1901 and 1907, each following approximately the same procedure: 'to buy, wait for an increase in value, and then sell'.[12] (Abu-Lughod, 1971, p. 153)

The incursion of foreign land speculators and investors as well as the introduction of new ownership laws and financial institutions trumpeted an up-coming boom in building construction. But for the rural areas, these changes, particularly in the land tenure system, did not mean that peasants would become suddenly wealthier, or that the real estate would flourish for the native Egyptians outside of Cairo. On the contrary, the changes in the rights of ownership, the increase in land tax rates, and the traditional inheritance system in Egypt, which tended to divide large properties to ever-smaller ones, made the great majority of landholders of agricultural land own less than five acres, hence, poorer (Zordoq, 1998, p. 364). On the other hand, a very small minority of Egyptian landowners held 44 per cent of the land, forming a new class, an agrarian bourgeoisie. Along with high officials, this minority constituted the Egyptian ruling class. Chichian reminds us not to confuse the emerging Egyptian bourgeoisie at the end of the nineteenth century with the European urban bourgeoisie, which was dependent on industrial production (Chichian, 1988, p. 33). In his analysis of the relationship between urbanism and colonialism, Anthony King has suggested that the shift to industrial capitalism at the colonial core is part of the same process as the shift to agricultural capitalism in the colonized periphery (King, 1990). It is important to add here a comment on how valid is Amy Chua's recent thesis in which she argues that free market economies do not spread wealth evenly throughout the whole society, but as Cairo's modern history tells us, produces a new class of wealthy individuals (Chua, 2002).

The concentration of economic surplus in the hands of a small, powerful and increasingly wealthy Egyptian bourgeoisie paralleled the concentration of the political, commercial, administrative, and leisure institutions in few urban localities around the country, which is a common feature of colonial domination. Cairo remained the dominant centre for this emerging network of urban centres. To connect these urban centres together, to Cairo, and to the outside world, the government extended throughout the country a network of roads, telegraphs, railways, and bridges. For governance and various services, skilled and unskilled labourers began to migrate to urban areas and, in particular, to the capital city. Besides the expansion of the government to accommodate the growing need for administrative services, the city drew an increasing number of landless peasants compelled to work as wage labour outside agriculture: from coal-heavers to domestic servants to transport and building construction workers (Beinin and Lockman, 1998).[13] As a result, during the 1882–1897 period Cairo had an average annual growth rate of more than 12 per cent (Chichian, 1988, p. 32). This excessive rate of growth to the city's native population paralleled an increase in its foreign inhabitants – in 1907 almost one of every seven persons living in Cairo was a foreign national (Chichian, 1988, p. 32). It is important to note that between 1897 and 1907, which is around the time Heliopolis was being conceived, the number of foreign residents in Cairo doubled. In addition to the permanent ethnoscape of the city residents, the beginning of the twentieth century witnessed a new, truly 'global' phenomenon: seasonal workers coming to Egypt every year from southern Europe to work in building construction during wintertime (Beinin and Lockman, 1998, p. 36).

The integration with the European economies facilitated the flows of Western goods and technologies, hence, altering the Egyptian technoscape and commodityscape, to use Appadurai's words. An enormous growth in imports of textiles and food was facilitated by the eagerness of the rising Egyptian bourgeoisie to mimick the European fashions and ways of life. It takes only the merest glance at issues of the daily newspaper, *Al Ahram*, from this period to recognize a striking increase in the number of advertisements for European modern gadgets: basic tools such as lathes, shapers, drills, and grinders, to other consumer items such as Singer sewing machines.[14] In building practices, not only building styles, but new materials and construction techniques using iron, steel, and reinforced concrete were introduced to Egypt a few years after they were used in Europe. For example, reinforced concrete was used in the construction of the Egyptian museum at the western edge of the Ismailia quarter only one decade after Francois Hennebique patented it.

The confluence of rapid urbanization, influx of foreigners, direct foreign investment, privatization, introduction of new technologies, and economic expansion exerted pressure on the physical structure of the city causing spillage beyond its peripheries. Roger Owen tells us that building construction prospered between 1897 and 1907. During this time, the Ministry of Public Works issued almost twenty thousand permissions for the construction of new buildings (Owen, 1970, pp. 509–510). As for the labour involved in this process, Owen tells us that in 1907 the number exceeded twenty-one thousand in Cairo alone. Building contractors were mainly Europeans from Italy, who had settled in Cairo since Khedive Ismail's construction boom of the 1860s and 1870s. Owen concludes that the prosperity of building construction, which was due to direct foreign investment and revenues from cotton exports, re-circulated most of the money within the country and it did not really leak outside, a view that is usually held (Owen, 1970, pp. 511–512).

But if there was a boom in construction and a pressure on the city to grow, to which direction did Cairo, actually, expand? From the founding of Al-Fustat in 640 AD, Cairo's urbanization has always tended to be along a south–north axis that runs parallel to both the course of the river Nile to the west and the Muqqatam hills to the east. The city expansion was therefore limited to the north and the south because of these natural barriers. It was not until the middle of the nineteenth century, and because of a slight recession of the river as well as filling and covering of some marshes in the area adjacent to it that the city expanded a little to the east. As we have seen, Ismail had converted this additional land to a modern urban commodity for sale. At the dawn of the twentieth century the city had new frontiers for its physical enlargement. Made possible by foreign direct investment, the building of the Aswan Dam in 1902 helped control the floods and flows of the river. This development facilitated not only the steady irrigation of agricultural lands along the Nile valley and the delta, but also the taming and confining of the riverbanks, particularly in the Cairo area. The stability of the river brought by the Dam had two important consequences on the urban growth of the city. First, it was possible to build bridges, hence, to cross the waterway and expand to the west; second, it became possible to fill and cover marshes, ponds, and swamps and, therefore, add more land to the city in the area between the river-proper and the *Muqattam* hills. Of course, the building of the Dam was parallel with other simultaneous developments that I have discussed in the preceding pages and other points I shall still discuss in the next section.

It was natural, therefore, with the increasing demand on housing that the city expanded in all directions peripheral to the north–south axis and across the river to the west. For the old, medieval area of Cairo – demographically described as the native and predominantly poor part of the city – the filling of ponds and marshes and the levelling of the rubbish mounds cleared some spaces for absorbing the incoming people from rural areas. However, the old city remained heavily populated and congested, and in the eyes of many, lacking the amenities of modern life. To the growing Egyptian middle class, the new bourgeoisie, and foreign nationals, the city needed new spaces. As I mentioned earlier, between 1898 and 1907, land speculation had grown immensely, pushing the prices of properties, particularly those located in the new Ismailia quarter, to record highs. Although the planning of the district and the construction of many buildings were done earlier, until the turn of the twentieth century there were still millions of square metres of vacant sites. The centrality of the quarter within the urban area of greater Cairo, the presence of modern amenities and entertainment facilities, and the prestige, and cultural capital that comes with being associated with the European way of life, made this area the most preferable to live in for upper class Egyptians. The other directions of development include the south, the north, and, after the finishing of the bridges, across the Nile to the islands of Zamalek, Rawda, and the Dukki area south of the village of Imbaba and north of the village of Giza. Abu-Lughod reports on three hundred land and building societies between 1901 and 1907 that worked in real estate development in these new areas. Naturally, the city expanded to become 108 square kilometres (Abu-Lughod, 1971, p. 122). Among the makers and shakers of the city's physical expansions at the beginning of the century was the Cairo Electric Railways and Heliopolis Oases Company, to which I shall now turn.[15]

Heliopolis: A European Vision of the Orient

With the clairvoyance of a successful businessman, Baron Édouard Empain, the Belgian banker and business tycoon bustled to Cairo in 1894 in anticipation of the economic boom. Armed with a fortune he had made earlier in the Congo and

France, he persuaded the government to grant him a concession to establish a tramway system in the city. By compressing space and time, the tram system changed the appearance of Cairo and the way of life for many Egyptians. The new European transportation technology revolutionized land speculation and speeded up the expansion of the city to the north, south, and across the Nile to Giza. Borrowing a simile, we may say that the laying of the tram rails was like drawing arrows that pointed at the direction of the city's future expansion. With every tramline laid, therefore, a fever of land speculation ensued. As one of the progenitors of the city's spatial expansion, Baron Empain recognized the power of his tram system, particularly the correlation between the laying of the tramlines and the increase in price of land around them. He also recognized that the success of any land development hinged on a well-established transportation system with the Ismailia quarter, the centre of the European district. With a shrewd perception, Empain wanted to hit two birds with one stone. His strategy comprised two steps: first, he wanted to control and monopolize the railway system in Egypt, and, second, to build a new city along one of its tracks. Samir Raafat tells us that in 1904, one area to the south between Cairo and Helwan and 'along the railway line that ran southwards from Bab al-Louk to Helwan' was his first choice (Raafat, 1996).

Failing to acquire the Cairo-Helwan line, Em-pain quietly shifted his search to the desert plateau north-east of Abassia – a stretch of land that was apparently far away from other expensive areas closer to the city centre, Ismailia quarter. In 1905, he established a joint stock company, the Ain Shams Oasis Electric Company with a capital of 15 million francs. Shortly afterwards, Empain and his Egyptian-Armenian partner, Boghos Nubar Pasha – the director of the Egyptian Railways between 1891 and 1898 and the son of a former Prime Minister – were granted two concessions from the Egyptian government. The first concession was granted to purchase almost six thousand feddans for the sole purpose of building a city; the second concession was to create a tramline for this new city and to connect it with Cairo.

Henry Ernest Jasper, the son of one of the principal architects of this new city, tells us that one day Baron Empain took his father for a horse ride across the north-eastern desert of Cairo, where the land designated for his real estate development was assigned. Empain exclaimed:

> Here, I shall build a city and call it Heliopolis, the City of the Sun; I shall start with a huge palace that I want to be magnificent (Heliopolis Palace hotel and now the presidential palace). I wish the architecture to be congruent with the traditions and customs of the land. I need a specialist in Arabic art, but I haven't found one yet. Do you like mosques? Aren't you an architect? Would you like to carry out my ideas? (Refaat, 1989)

What were these ideas, and how were they implemented in the new development? Why did Empain want the Arab style in his new city (figure 7.1) ? And if he admired so much the Arab architecture of the mosques, why did he build his own palace later in a Hindi style? Once again, there are many strands to the Heliopolis story. Each one could be pursued in our investigation of how this particular real estate development was linked to the 'global' forces of its time and how it reflected this growing fantasy with the orient. I want to pick up a particular strand and proceed by commenting on an observation Robert Ilbert has made concerning the original source of design and character of the new development, Ain Shams, or later Heliopolis (Ilbert, 1981, 1984).

In his seminal work on Heliopolis, Ilbert asserts that the land development was designed according to the Garden City theory – a theory of city

Figure 7.1. Heliopolis, a pastiche of an European vision of the orient put in stone.

planning advanced by Ebenezer Howard at the end of the nineteenth century. Writing against the background of the industrial depression and growing labour unrest that engulfed Europe during the third quarter of the nineteenth century, Howard developed a sanguine view of a utopian society in which town and country were united against a vitiated city and severe agricultural depression in the countryside. The influx of rural dwellers moving to the overcrowded cities, increasingly aggravating the urban situation. His antagonism towards the city emanated from his identification of it with extreme concentration of wealth and power as well as poverty and misery. For him, these extremes characterized the modern industrial city in Europe with the complementary features of 'palatial edifices and fearful slums' (Howard, 1902), to use his words. Howard believed in decentralizing the city. In his view, while steam energy and railways had brought large factories and cities, the dawning age of electricity would make possible a rapid decentralization of the urban population. In his lectures and writings, Howard expounded the different socio-political and economic aspects of decentralization via developing small cooperative communities which he strove to promote as magnets for drawing people, particularly the working class, away from the congested urban centres. In searching for the essential components of his new town-country magnet, Howard sought to marry the benefits of town and country: the social, employment and amusement opportunities of the town and beauty, wealth, freedom and clean air and water of the country. The former was a mix of spaces for consumption and amusement whereas the latter was reserved for the spaces of the highest values for the community, namely, cultural spaces. It is important to mention that Howard was formulating his theories at a time the two ideologies of capitalism and Marxism were at murky loggerheads in their competition for dominating a world vision. While Marxism may have influenced him, Howard strove to forge a third way that would amend capitalism without doing away with it. Of course, we know historically that Howard's Garden City theory was discussed widely at around the time Heliopolis was being conceived; hence, there is no doubt that Empain, or one of his principal architects may have come across Howard's writings and ideas.

But how did the ideoscape of the Garden City movement and other prevailing ideas at the turn of the century influence the planning of Heliopolis? Was Heliopolis meant to be a cooperative community? Ilbert tells us that the similarities between the original design scheme of Heliopolis and that of Parker and Unwin's plan for Letchworth in 1903 are striking. For instance, an echo of Parker and Unwin's geometrical formality, which they emulated from Howard's diagrams, was retained around the main square of Heliopolis, where later the main cathedral was located. In addition, the area of Heliopolis, its capacity of inhabitants, and the selection of its location far from Cairo may have been influenced by the panacea of the urban ills of cities promoted by Howard. While the Garden City was supposed to occupy 1000 acres (405 hectares) surrounded by a much larger area of green belt with not only farms, but institutions such as convalescent homes which would benefit from a rural location, giving a total of 5000 acres or 2235 hectares, Heliopolis was two thousand five hundred, expandable to five thousand, acres with only one-sixth of its total area designated for urban development. It is worthwhile adding that at that time the area of Heliopolis was almost as big as the whole city of Cairo. But the benevolent, cooperative approach of Howard was perhaps hardly convincing to someone with Empain's strong capitalist inclinations. For Howard the cooperative methods of the Garden City 'will by a purely natural process, make it gradually impossible for any landlord class to exist at all'; (Howard, quoted in Fishman, 1982) hence, all the people of the city would enjoy the privileges of profit making. For Empain, Heliopolis was no more than a capitalist venture, a real estate development in a foreign land with an initial capital investment of 15 million francs that reached 50 million around 1911; at that time, Heliopolis was the second biggest foreign investment in Egypt after the Suez Canal. While Heliopolis may have had formal planning similarities with the Garden City, it did not share Howard's goals and socio-political intentions. My point here is that Empain had managed to strip the socialist, cooperative agenda out of the concept of the Garden City and re-invented it solely in formalistic and marketable terms. Put in this way, I want to suggest that this attitude was essential to creating a fantasy environment, a point I shall pick up later.

The initial capital investment in Heliopolis was unmatched by any of its contemporary real estate developments and, certainly, drawing people to the desert was a risky business for such a huge sum of money. For its progenitors, it was essential, therefore, to develop a strategy for drawing people to the desert and creating a successful real estate development that would guarantee a decent return on investment to the Belgian stockholders. Of course, resorting to the Garden City planning ideas was not enough reason to convince people to move out of Cairo to the surrounding desert; nor was it clear why these ideas were attractive to a capitalist like Empain. True, following what Howard propagated, electricity and the tramline were thought of as major factors to draw people far from the city centre by facilitating a speedy transportation system – that was what Empain had done after all. What needed to be determined was the segment of Cairenes in the consumer market at the time who had the purchasing power that Empain wanted to draw to his new city?

Before I begin addressing this question, I want to take a note of two important facts. First, Empain's approach to building Heliopolis was to create a total environment: the new development was meant to be independent of Cairo, yet linked to it. Second, it was the first time in its history that Cairo expanded to the surrounding desert, far from the agricultural lands around the Nile. This fact created a condition of uncertainty about the project, which is why it undoubtedly required the development of a new marketing strategy. The most viable targets for the project were the neo-Egyptian bourgeoisie and the foreigners residing in or moving to Cairo. However, the market share from these segments was relatively small, particularly when we take into consideration the number of real estate development projects going on at the same time. Another growing segment of Egyptian bureaucrats, the government employees, was a potential target even though it was apparently more convenient for these people to reside close to the government buildings that surrounded the Ismailia district. As for the Egyptian working class, they always lived within the medieval city and, needless to say, their purchasing power was insignificant for any profit making. Obviously, with the exception of the working class, all of the other segments were potential targets for Empain. There was another growing segment, however, which Empain had in mind, and its requirements were essential factors, I shall argue, in deciding on the planning approach of Heliopolis as well as its architectural style, particularly in the first stage of the project before the Cairo stock market crash in 1907.

The growing tourism industry indeed was responsible for the multiplication of the foreign presence in Cairo. Thomas Cook, the founder of the travel agency bearing his name, capitalized on the growing European fantasy of the chimerical Orient and the increasingly cheap travel fares around the Mediterranean and began organizing excursions to visit Egypt as early as the 1860s. Of course other forces in fantasizing and 'orientalizing' the Orient include world exhibitions, which were popular in the second half of the nineteenth century. These tremendous 'international' fairs introduced images of 'oriental' Cairo to many Europeans, manifesting an oriental 'phantasmagoria', to use Walter Benjamin's word. Whether from writings, paintings, exhibitions, or travelogues, inducement of an imagined exoticism of the orient for Europeans fashioned a dreamscape and a desire to travel and immerse oneself in the East's allure. At that time, the tourism industry in Egypt was based upon the ancient heritage along the river Nile and the great pyramids of Giza as well as the mosques, bazaars, and *madrassas* of medieval Cairo. With the subtle economic, political, and social changes in Western Europe, particularly the emergence of a leisure class, the tide was in favour of expanding tourism in the Middle East along with its integral, essential components, hotels. During these days, the Ismailia quarter, particularly its eastern edge overlooking the Azbakiya gardens, was the main area in the city where famous hotels were located and from which foreign tourists gazed out over 'oriental' Cairo. The Shepheard Hotel, probably the most celebrated hotel during the colonial era, had a panorama from its terrace that probably never satiated even the most blasé orientalist. Scenes from an 'oriental' way of life were passing below the terrace against which the silhouette of minarets and domes of medieval Cairo could be seen.[16] The architectural style of the hotel was very European. This style and its location overlooking old Cairo reflected the tourists' desire for authentic, exotic experiences while retaining the amenities of modern life: seeing the other culture while keeping a comfortable distance. The terrace of the hotel both symbolized and reconciled this dual desire perhaps more than anything else. In fact, the tourism industry until today plays on this duality.

I suggest that Empain wanted to establish his city as another tourist destination and his 'hotel palace' as another Shepheard Hotel. But why would tourists go to Heliopolis? We learn from Anne van Loo that Empain was in contact with the Belgian archeological mission in Egypt and must have been influenced by it to believe that the land he had purchased was where the ancient city of 'On' had once stood (quoted in Hassan, 2001, p. 16). On, Heliopolis, or the City of the Sun, as it was known in history books, was a good founding myth for his new town as well as for attracting tourists to his land. Thus, immediately after purchasing the land, he ordered excavations on the site. The effort to discover On, of course, did not yield any findings because the actual site of the ancient city turned out to be outside his land, more towards the delta and the village of Zeitoun. Since the land purchased was 'historically' insignificant, the selected character of the buildings, like the palace hotel, depicted what may seem to mimic an Arabian history. In choosing the Arabian style for the hotel, Empain wanted, therefore, to complement the intended Oriental atmosphere of the place. In addition to the architectural style, the connotation of 'an oasis', encapsulated in the marketing of Heliopolis, projected a stereotype of the Orient with which Europeans were familiar. To be sure, the image of an oasis must have had resonance of titillating scenes and imageries from 'A Thousand and One Nights': Bedouins, harem, haunting memories of romance, Flaubert and Kuchuk Hanem, and a whole exotic, Oriental lifestyle. The Oasis, therefore, seemed to belong to what Edward Said described as 'a system of representations framed by a whole set of factors that brought the orient into Western learning, Western consciousness, and later, Western empire' (Said, 1979, pp. 202–203). Edward Said has shown us that Orientalism in the nineteenth century has been an indispensable concept for any understanding of the power to colonize. Similarly, Timothy Mitchell has contended that colonization referred to the inception of new forms of order, among them a new spatial organization and planning of cities (Mitchell, 1991).

Under colonialism, old cities and new ones were developing according to a colonial order that was part and parcel of the expanding capitalist order (King, 1990; Wright, 1987). In a comprehensive survey of literature on colonialism, Telkamp has indicated some features that characterize cities developing under colonial rule, of which I find some applicable to Heliopolis (Telkamp, 1978). For example, he posits that Western models of urban design dictated the urban forms of colonial cities. I have indicated already that the planning of Heliopolis mimicked the concept of the Garden City in its urban organization. Moreover, Telkamp points out that any colonial city exhibits some sort of residential segregation between exogenous elite and indigenous inhabitants, and this, too, is true of Heliopolis. Unlike Howard's conception of the Garden City where boundary lines between classes were abolished, the initial planning concept of Heliopolis included two distinct oases separated by a 1 kilometre strip of desert. While the first oasis was reserved for foreign residents, tourists, and upper class Egyptians, the second oasis was designated for the natives, who would form the working community for the other oasis. Significantly, the first oasis was organized around a cathedral and the palace hotel whereas the second oasis was organized around a mosque. It is worthwhile noting that the initial strategy for planning the development was to construct, over a long period of time, a series of oases.

We should also keep in mind that the planning of Heliopolis was indirectly influenced by the introduction of new European medical theories, particularly the miasmic theory – that disease resulted from polluted air (King, 1990; Mitchell, 1991). Besides projecting a sharp contrast to the crowded, unhealthy, and noisy Cairo, hence, provoking feelings of green, crisp air, and tranquil ideal environment for attracting Cairenes residents, the new oasis was planned to develop as low-density housing, separated from the native polluted city. We should bear in mind that until the time of Heliopolis, Cairo did not have a sewage system; hence, it had yet to get rid of its epidemic diseases and plagues. The most recent plague had struck a few years earlier, in the summer of 1902, when an outbreak of cholera hit the city resulting in a death rate of a thousand a day.

It was not a cholera plague that slowed down significantly the construction fever in Cairo, but a stock market crash that took place in 1907. Abu Lughod describes the financescape during these days:

> Recognition that credit had been overextended, coupled with the general tightening of money that was later to result in worldwide panic selling and bank runs, led to the abrupt suspension of Egyptian credit in both Paris and London. The effects of this sudden prick upon the speculative bubble were as immediate as they were catastrophic. (Abu Lughod, 1971, p. 153)

If, hypothetically, a contemporary Cairenes travelled back in time to these days, he or she would have found a very similar scene to today's cityscape: 'a wilderness of half-finished buildings in Cairo rivalling that of the archeological remains of upper Egypt' (Cunningham, 1912, quoted in Abu-Lughod, 1971, p. 138). Many speculators went bankrupt and scores of real estate developments were abandoned. Due to these developments, Empain was forced to change his ambitious plans: the intended series of oases was reduced to one, namely, Heliopolis oasis. Interestingly, the Heliopolis oasis remained divided internally, as in the former plan, between the residences of foreigners and of natives. For the company, other means of attracting people were adopted and new negotiations with the government transformed the initial plan. For example, the company struck a deal with the government to build apartments for the growing white-collar government employees, thus, securing itself a certain number of residences.

But one significant development to the planning of Heliopolis in the aftermath of the stock market crash was that the new scheme included leisure and entertainment spaces on a scale previously unprecedented in Egypt: a Luna Park, hippodrome, aerodrome, cricket, polo, and golf courses, and a grand Brasserie Restaurant. By 1915, the entertainment and leisure spaces, the essential components of Howard's third magnet, comprised 25 per cent of the total built area. Thus the development was becoming more a place for play and fun and less as a place for work. It could be said that prior to the introduction of entertainment activities and commercialized leisure in Heliopolis there was no public entertainment zone in the whole of Cairo. In fact, following the opening of the Heliopolis palace hotel, the fast metro tram, and the various entertainment places, Heliopolis became the entertainment hub of Cairo. These entertainment spaces reflected the fledgling new culture of entertainment in Egypt, which was part and parcel of the emerging Egyptian capitalism and a European-bourgeois way of life. In a way, the emergence of this urban pattern of leisure can be attributed to the changes in the socio-economic spheres, particularly the changes in the ethnoscape, technoscape, financescape, commodityscape, ideoscape, and mediascape. In addition, in sharp contrast to the disorganized reality of the old streets of medieval Cairo and the Arab architecture, Heliopolis was a measured, controlled, and organized urban experience, which was intimately linked to a fusion of strategies of entertainment, orientalism, tourism, and capitalism. With our contemporary eyes, Heliopolis was a postmodern development. I shall return to this point again. Now, however, I would like to move forward in time, to the last decade of the twentieth century and look at the more recent condition of Cairo.

The Postcolonial Economy and Mixed Capitalism

The year 1991 is one many commentators and analysts of Cairo's modern history will remember for years to come. Interrelated national and 'global' political and economic events that year have led to significant developments in Egypt's socio-political and economic spheres as well as the urban condition in its capital city. One incisive event was the decision taken by the Egyptian government to join the American-led coalition in the second Gulf War, which won the country a brief economic respite in the form of writing off good portions of its debts to both the United States and the Gulf States. Moreover, the government's political position during the war pleased the World Bank and the International Monetary Fund and set the general mood in both the government and the international agencies for putting Egypt onto 'an economic reform path', which the country has since then been traversing. It is worth mentioning that the reform path, known as the Structural Adjustment Program (SAP), came almost a century after the privatization fever that hit the country at the end of the nineteenth century. As in the old days, the government embarked on a round of selling off public assets in a much-debated privatization and *laissez faire* policy. Between then and now, this turbulent and momentous century had left undisputed marks on the urban spaces of the capital city. These tumultuous years – in particular to our interest here the last decade of the twentieth century – have unleashed new socio-economic and political forces upon Cairo, to which the city has responded physically by expanding in every direction and in ways that are simultaneously old and unique. To better understand the recent development, I shall put these forces in their socio-political and economic contexts; hence, I shall very briefly trace their roots in other earlier significant dates and historical events, particularly those that affected directly the urban condition of Cairo during the 1990s.

Without any doubt, one of the most significant events in Cairo's modern history is 23 July 1952. Egypt's postcolonial history began on this date when a *coup d'état*, led by Gamal Abd El-Nasser, ended the monarchy and, shortly afterwards, led to the declaration of Egypt as a republic. Although the regime of the Free Officers, as they were called, began as a nationalist revolution with no clear ideology, it soon began shaping its programme around some socialist ideas. (Egyptian socialism has commonly meant the centralization of government and state ownership and management of resources and means of production.) If we believe Henri Lefebvre that any revolution which has not produced new spaces 'has not changed life itself, but has merely changed ideological superstructures, institutions or political apparatuses' (Lefebvre, 1991), we should then ask the following question: what kind of spaces has the 1952 Revolution produced in Cairo?

To a great extent, the new spaces produced in the years immediately following the Revolution reflected the nationalist aspirations of the government. The Nile Corniche, or shore drive, was one fine example of a space produced by the government as a showcase of national pride. Prior to the Revolution, the Nile shores, particularly west and south-west of the Ismailia area, were almost exclusive to the private palaces or the British embassy and army barracks. From the new regime's point of view, carving out this space along the Nile not only freed spaces from the grip of the former 'colonial' power and its Egyptian bourgeoisie collaborators but also reconfigured them as signs of a new era. Of course, these spaces became prime properties for development, which marked the new postcolonial city. Indeed, the area that once separated the Ismailia quarter from the native city during the colonial era and was regarded as the city's finest real estate, was eclipsed during the postcolonial spatial upsurge with the land facing the Nile and close to Tahrir Square. I shall propose that the rebuilding, in 1957, of the famous Shepheard Hotel along the new Corniche, after it was burnt during the Black Saturday riots that preceded the 1952 Revolution, symbolized the passing of colonialism and Orientalism and the emergence of postcolonial time and spaces.

But the postcolonial spaces that most notably came to affect the city in the aftermath of the Revolution were those reflecting the government's ideological ambitions at the northern and southern edges of the metropolis. We know that the revolutionary officers came with loose ideas about how to reverse the feudal order of the former monarchy and the imperial order of the Western powers. Their ambitions included, among other things, modernizing the country and abolishing the class differences. At that time, modernizing the country was seen as synonymous to industrializing it so that Egypt could equal the West on its own terms. Thus we find the development of industries inside and around Cairo, particularly in the north and south of the existing metropolis, characterizing most of the spatial expansions during the 1950s and 1960s.[17] Thus, at around the end of the 1960s, the industrial zones in Cairo comprised about 13.5 per cent of the total land use in the city. The establishment of these industries and their corollaries of massive housing projects for the newly-arrived rural migrants, who constituted most of the labour force, blocked the horizontal north-south pattern of city growth, which until then characterized how the middle and upper classes had always expanded historically.[18] Thus the spatial expansion of the elite class began to take place vertically rather than horizontally and in specific parts of the city, like Heliopolis, Zamalek, Garden City. It is therefore hardly accidental that the year 1958 witnessed the construction of the first high-rise building in Cairo, in the district of Garden City, known as 'Belmont Building'. Certainly, there were other horizontal expansions during this era, particularly for the expanding middle class, such as Nasr City, north-east of Cairo. Nasr City was planned as the government's new centre to be populated by its bureaucrats and civil servants.

A second point to be remembered from the years following the Revolution that reflected its ideology and which also had a tremendous impact on the real estate market until today, was the freezing of apartment rents as a means of price control and wealth redistribution. Between 1952 and 1974, one housing legislation after another aimed at reducing the rental rates and making it virtually impossible for landlords to evict tenants when their contracts expired. The result was catastrophic to the environment and to the real estate capital market. First, even today, tenants are still wedded to the idea inherited from these days that once they have moved into a rented apartment, it is theirs for life. Many owners would therefore prefer to keep their apartments closed rather than risking renting to tenants who may later refuse to leave. Press reports today estimate empty apartments in Cairo alone at between 14 and 18 per cent of the total number of units. Second, although inflation was on the rise, the rent remained almost fixed and, therefore, landlords could not afford maintenance and most buildings were left to deteriorate. Third, coupled with these two previous points, housing legislation during this period made it unattractive for the private sector to participate in real estate ventures and, therefore, the government took full responsibility for meeting an ever-increasing demand for accommodation. But, because the population was rising faster than the rate of building new housing units, let alone the impediment of the political atmosphere in the Middle East, which resulted in military and defence needs absorbing most of the country's resources, the government could not afford the tremendous task it set itself to achieve. As a result, the informal housing market flourished, particularly illegal construction and squatting. To add to the woes of the housing shortage as well as private enterprise altogether, the regime made its most aggressive attacks on the former landed aristocracies and big capitalists by sequestrating their lands and businesses. These policies not only discouraged all national private initiatives in housing or in any other private business, but also prompted foreign nationals to leave the country along with foreign investment, foreign commodities and foreign technologies, putting a dramatic finality to the liberal days of the early twentieth century. This chapter in Egypt's modern history was, however, abruptly relinquished after the death of Nasser and the rise of his successor, Anwar Sadat in 1970.

Four years after consolidating power and one year after completing a cease fire agreement with Israel following the 1973 October War, Sadat introduced a major policy change, known as *al-infitah* (opening of Egyptian economic market to the West). Arguably, *al-infitah* ushered a new round of economic and political liberalization, a new opening to the international economic environment. Whether viewed in positive or negative terms, the visible aspects of this policy included a relative increase in foreign commercial banks and private investment companies and a boom in the construction sector, particularly for rebuilding the canal cities and new luxury apartment buildings and hotels in Cairo. Saad El-Din Ibrahim draws parallels between Sadat's vision for Egypt and Cairo and Khedive Ismail's a century earlier. In a much-celebrated quotation Ibrahim contends that Sadat 'wanted to develop Egypt along Western ideals and fundamentals, with Western money, technology and experts. If Paris and Rome were favourite models for Ismail, Los Angeles and Houston were favourite models for Sadat' (Ibrahim, 1984, p. 56).

From a different perspective, this revival of Egyptian capitalism benefited many of the former bourgeoisie as well as a nouveau riche class of businessmen – both beneficiaries invested their money in land, apartments, and various imported consumer items. It is worthwhile highlighting that in the post-*infitah* years, the country imported Western technologies, commodities, and ways of life and exported, besides petroleum to Europe, human labour in search for higher wages to rich Gulf States. The *infitah* did not come near a total transition to a free market capitalist system. The

Egyptian economy remained isolated from the international economic markets because of its very large, sluggish, complicated economic systems, dominated more by the public sector and central planning. Moreover, direct foreign investment did not come 'anywhere near the numbers that Sadat had hoped for' (Goldschmidt, 1988).

Many economists, therefore, attribute the economic revival of the 1970s not to the *infitah* policy *per se*, but rather to the significant increase in the oil production and prices, the reopening of the Suez Canal, tourism, and remittances from Egyptians working on the Gulf. (El-Nagar, 1997) To put this economic growth in numbers, the average growth in GDP between 1974 and 1984 was 9 per cent per annum (Handoussa, 1991). But the waning of the oil boom of the 1970s during the 1980s revealed some tremendous economic problems for the Egyptian economy, which was already declining, with the GDP indicators dropping and the population, inflation, budget deficit, foreign debts, and unemployment rates soaring. The Egyptian government subsequently approached the World Bank and the International Monetary Fund (IMF) in 1987 to help ameliorate the burdens of the accumulating foreign debts in return for some limited economic reforms. These changes did not stop the stumbling economic situation from getting worse, a condition that remained until June 1991, when the agreement between the Egyptian government and the World Bank and IMF was reached. The agreement meant that Egypt would be provided with billions of dollars in loans on the condition that the government agreed to undergo a structural adjustment programme (SAP). The standard economic procedure of the programme included three basic tenets for the Egyptian government to undertake: reorienting the economy towards exports to earn foreign currency required for servicing the debts; deregulating the financial sector so capital can flow freely across the national border; and privatizing the public enterprise. Of course some of these reforms and measures had already been legislated earlier during the heyday of the *infitah*.[19]

With hindsight, what was taking place during this historical juncture, in the aftermath of the collapse of the Soviet Union, can be described as part of a global restructuring process, a period that marked a critical turning point in the geohistory of capitalism. It was a period that witnessed the rise of the neoliberal paradigm to an almost hegemonic status. In this dramatic, staged, comeback of core capital as a sole, dominant, and centralized global power, the World Bank and the IMF, like the British counsul-general a century earlier, became the principal tools through which the ideology of neoliberalism has been promoted and policed (Alam, 2003; Amin, 1997).[20] By and large, neoliberalism is a wide-range ensemble of economic, political, social, and ideological discourses and strategies, which combine aspects of liberal and conservative politics. Its main economic tenets revolve around the free market society and *laissez-faire* politics (Overbeek and Pijl, 1993).[21] Although its origin cannot be traced back to any one single incident, it is apparent that it developed in the early 1970s, at around the time the international monetary system was breaking down. Since then, the global capitalist system, led by the United States of America and Britain, has been increasingly neoliberalized. During the late 1970s and the 1980s, neoliberalism had been promoted vigorously by the conservative Thatcher and Reagan administrations. By the mid 1980s, neolibral economic policies took another thrust due to the revolution in communication technologies, which has brought radical changes in the speed and scale of all financial activities. The expansion of abstract financial speculation as well as cultural consumption and the service sector coincided with the rise of what Sharon Zukin has termed the cities' 'symbolic economy' (Zukin, 1995). Zukin has contended that the symbolic economy, the blending of cultural symbols and entrepreneurial capital, in advanced capitalist countries, has significantly privatized a great number of public spaces. I want to extend her observation to the context of Cairo, and suggest that the current global forces of neoliberal economies are increasingly commodifying and privatizing spaces for cultural consumption. I shall return to this point shortly.

Against this brief overview of global capitalist financescape in the past few decades, I want to proceed by emphasizing that the underlying neoliberal economic policies implemented, or rather, domesticated, in Egypt during the same time, particularly in the 1990s, have had a profound influence on Cairo's cityscape.[22] Of course, this is a very large and thorny issue, and, therefore, I am going to touch only its edges. Because it is a controversial issue, I want to begin approaching it by focusing on the feature that appears in the writings of most commentators on today's Egypt, namely, the disparities and inequalities that exist in the society at large as a consequence of the recent economic developments. True, compared with the troubled days of the late 1980s, the macroeconomic indicators' speed of turn around during the 1990s 'qualifies as one of the most impressive in recent economic experience anywhere in the world' (Ikram, 2001).

But, I still want to emphasize the unevenness and inequalities of the wealth distribution and the ramification of this condition on the cityscape. Recall Amy Chau's thesis: in all countries that adopted the free market ideology, wealth was not spread evenly throughout the whole society, but was concentrated on a new class of wealthy individuals (Chua, 2002). Similar observations have been made in numerous critical writings and analyses of globalization. For instance, Walden Bello discusses the social disparities created by the implementation of SAP in various countries in the world (Bello, 1996). In terms specific to Egypt, Timothy Mitchell argues that the reform plan concentrated the public funds into fewer hands of 'ever more powerful and prosperous financiers and entrepreneurs' (Mitchell, 2002, p. 282). Mitchell quotes the Economist Intelligence Unit quoting the 1994 United Nation's Human Rights Report which recognized Egypt as 'in danger of joining the world's list of failed states because of wide income gaps between sections of their population' (Economist Intelligence Unit, 1996/97). This condition is particularly glaring in Cairo where the upper class continues to be conspicuous (Yousry *et al.*, 1998; Gauch, 2000). These disparities and income gaps have been elucidated recently by Rushdi Said. In a book entitled *The Truth and Delusion of the Egyptian Current Condition*,[23] Said categorizes Egyptian society as consisting of two sections, which he terms the submerged and the floating sections – of course, in relation to poverty line (Said, 1996). He estimates that 86 per cent of the Egyptian society belong to the submerged section, which live in dismal conditions and receive only about 26 per cent of the GDP. Said divides the remaining 14 per cent of the floating section internally to three sub-sections. At the top is a group of wealthy Egyptians that constitute nearly 2 per cent of the population. This segment, which receives 40 per cent of the GDP, is 'exceptionally and often ostentatiously wealthy', to use the language of the US Embassy in Cairo report (Economic Intelligence Unit, 1969/97, p.22). At the bottom of this section lies about 6 per cent of the population, which receives 9 per cent of the GDP and is struggling to keep afloat. Between the two preceding segments, argues Said, there is a third one that constitutes 6 per cent of the population, and receives about 25 per cent of the GDP.

Other official numbers are less grim than Said's accounts. For instance, taking the percentages of housing types built between 1982 and 1997 as an indication of the social class distribution, a recent study found that 66 per cent of the urban housing belonged to the low-income group, 24 per cent to the middle-income group, and 10 per cent to the high-income and luxury segment (Ministry of Planning, 1997). Whether one takes Said's or the official account, these nation-wide rough figures are indicative of the condition in the Greater Cairo Region (GCR), an area that is home for almost one in every four Egyptians, and whose population increases by 3 per cent every year.

In tandem with the privatization and liberalization modes of thinking and *zeitgeist*, particularly because of the, arguable, economic success, the government started reversing many of the laws inherited from the Nasserite, socialist era. For example, although it was only a small step towards full liberalization of the housing market, the parliament recently passed an amendment to the rent law, which allowed the renegotiation of rents. Other steps include drafting a new mortgage law and easing restrictions on foreign ownership of land. Furthermore, the reform steps imposed by the IMF and the World Bank spurred the government to change further its function and involvement in the housing industry. For example, the public sector's involvement in the production of houses dropped to 5 per cent of the total housing units built between 1982 and 1997 (Ministry of Planning, 1997). It is important to note that since the early 1980s, the state has only focused on producing and subsidizing houses for the low-income groups.[24] But of course subsidies and the social welfare policies contradicted the neoliberal mantra of free markets and privatization. To resolve this contradiction, the state had to cut the budget deficit by reducing all housing loans and subsidies while at the same time increasing the interest rates on the subsidized loans for housing, from 4 per cent to 6 per cent with larger down-payments (Muselhy, 2002). In a matter of a few years, therefore, many people from the low-income segment of the population, the submerged section, found the formal, subsidized economic housing beyond their reach. Consequently, a study conducted in 1993 found that 52 per cent of the population of Greater Cairo Region (GCR), around five and a half million, live in informal, squatted, degraded areas, comprising around 92 square kilometres (IDMSC, 1993). Most of these areas are scattered either inside, or at the administrative boundaries of, GCR, where ownership of land is disputed among various governmental bodies and governorates. It is estimated that these marginal areas – usually developed as encroachments on the agricultural land surrounding the metropolis – are increasing by 150 feddan every year. Moreover, recent studies reveal that many inhabitants of these informal, urban, marginal areas are not migrants from the countryside, but rather educated professionals and governmental employees, who moved from within the city (Hanna, 1996).

Inner-city slums, squatter settlements, and all types of substandard, informal housing have been aggravated further because of the effect of the economic reform plans on the local currency. Since the days of *infitah*, the value of the Egyptian pound has been constantly falling, with the actual wages of labourers and government employees falling sharply. Indeed, the average family's purchasing power declined significantly, giving an indirect blow to the housing market. Naturally, the lack of adequate, affordable housing forced many newly weds and families to live in alternative spaces: graveyards, tents, workshops, garages, riverboats, staircase vaults, shared rooms in apartments, and rooftop shacks. The dismal living conditions of many inhabitants, directly or indirectly, put a tremendous pressure on the city to expand. This is so because the impoverished, degraded environments and the demographic explosion of the GCR – caused by natural growth, rural migration, or expansion of GCR's administrative boundary – increased the pollution of the city to alarming levels. Besides the visual cacophony of deteriorating buildings and clashing architectural styles, colours, and height, the immediate environment has been increasingly degraded: high urban noise level, streets littered with refuse, blackish smoke hanging over the city, and never-ending traffic jams.

No part of GCR, rich or poor, has been immune to the effects of pollution and demographic explosion. For the upper middle and upper classes, this condition resurrected the idea, which Baron Empain started at the beginning of the century, to leave the city for the surrounding deserts. In this regard, the state has been supportive for two main reasons. First, by directing the city growth to new settlements and development axes, the living environment of the GCR would improve. Second, the selling of desert land opened up new ways of profiteering for the government and a new class of entrepreneurs that is increasingly linked to both the state and the international corporations. The selling-off of some public land accommodated a massive transfer of the desert surrounding Cairo to private ownership. Let us not forget that the neoliberal economic orientation already formed a background that supported the privatization and commodification of land. The Minister of Housing echoed this spirit of deregulation and privatization. In a televised interview a few years ago, Mohamed Solieman pronounced that 'land constitutes the assets of any country'. Against the conventional wisdom of regarding people as the assets of nations, the remarks reflected the government's priority in expanding and supporting the real estate market. To put these new real estate expansions into numbers, Muselhy writes that the total area of GCR in 1977 was about 50,000 feddans (Muselhy, 1988). According to a recent governmental publication, the total urban area that has been added to the GCR, since Law 59 of 1979 on New Urban Communities, is about 175,000 feddans – this figure does not take into account the squatting and informal expansions of older areas (Ministry of Housing, 2002). Today, the urban area of GCR is, therefore, more than a quarter of a million feddans in a total area of 850,000 feddans (figure 7.2). To put it differently, in twenty-five years, the area of GCR has more than quadrupled. What is more striking than this vast unprecedented urban expansion is that large areas of it are exclusively, privatized public spaces for the elite, ostentatious segment of society, mirroring the disparities and inequalities that exist in society. It is to these privatized, fantasy spaces of consumption that I shall now turn.

Dreamland: An Oriental Vision of the Occident

Like Baron Empain, Ahmed Bahgat, the Chairman of the Bahgat Group and the progenitor of the huge real estate project, Dreamland, is an industrialist who, along with many other Egyptian entrepreneurs, jumped on the bandwagon of real estate business during the economic boom of the 1990s.[24] The anticipated quick, high profitability of land development in comparison with other manufacturing businesses was the lure to enter the real estate market. According to the Investment Authority, between 1993 and 1999 the numbers of real estate developers have more than tripled to reach ninety-two with a capital of 3.3 billion Egyptian pounds (Mekay, 1999). Although specific numbers are not available, it is certain

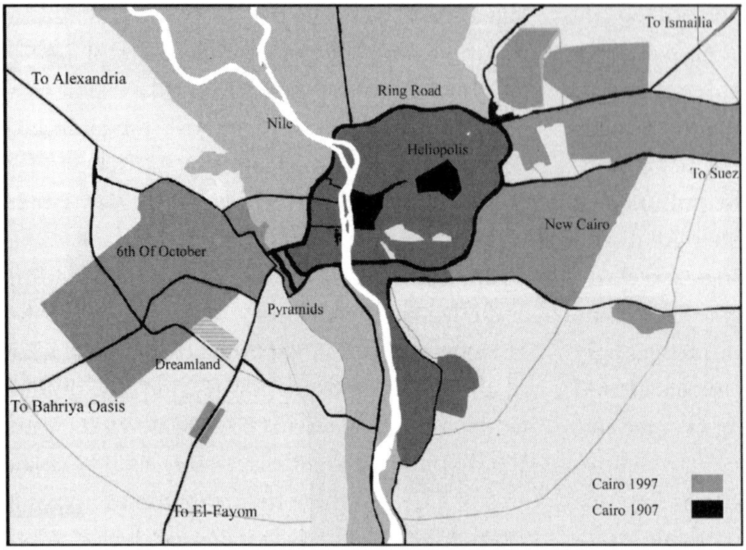

Figure 7.2. During the course of the twentieth century Cairo's urban area has increased more than twenty-fold.

that it is Egyptian, not European, industrialists, who are investing extensively in the leisure, upmarket residential communities and own a large percentage of these companies. In a matter of a few years, a new type of urban development – the gated community – became a mass trend prevalent in Cairo's ever-expanding cityscape.

Like a walled oasis, the first rudimentary prototype of residential gated communities in Egypt appeared in the early 1960s as a holiday resort-village, Al-Ma'amura, east of Alexandria – now part of Alexandria. Financed and built by the state-owned company bearing the same name, al-Ma'amura was a semi-public development, and most of the people who stayed there were temporary summer vacation residents. Moreover, its beaches were public and access was obtained with a small ticket-price. It was not until the 1980s that resort-villages were rapidly built along the Mediterranean coast, west of Alexandria. Most of these new resorts differ significantly from Al-Ma'amura, particularly in their exclusiveness to those who can afford to buy summerhouses built there. For residents of these summerhouses, these resorts provided new enclosed spaces for their conspicuous consumption. But for many other Egyptians, like those expatriate workers in the Persian Gulf, who earn relatively high wages, there were two further motivations for investing in this type of development: speculation and boom in leisure and entertainment industries. High inflation rates during the 1980s made it more profitable to invest in real estate, summerhouses, and commodities in general than to save money in banks.

In the early 1990s, a desire to extend the summer resorts' leisure time and spaces to Cairo grew fast: if the lifestyle of the summer resort was appealing to many elite Cairenes, why not extend this type to Cairo all year long? In the meantime, the state was busy deregulating not only the country's financescape, but also its mediascape. (I want to suggest that the deregulation of the media played a crucial role akin to that carried out by nineteenth-century writers, painters, and World Fairs.[26]) True that since the days of *infitah*, not only tangible Western commodities such as Coca-Cola and Levi jeans had become familiar to all Egyptians, but also images and scenes of Madonna, Dallas, and other North American media personalities and creations were popular in certain circles. During the 1990s, these images exploded upon the population at large. In this regard, I am far from claiming that these images are creating a homogeneous culture. Rather, I would like to emphasize that the global distribution of North American images to Egyptians through the media facilitated the introduction and the promotion of the concept of lifestyle, which depicts particular desires and tendencies taking place as a result of globalization. To be sure, while many of the international brand commodities that circulate with particular lifestyles are not available for all Egyptians, these items certainly became present in their lives as images interwoven with particular ways of life. Most importantly, these images of the 'American way of life' were interwoven with their global spatial manifestations and building types such as malls, theme parks, fast food restaurants, and gated communities. Today, these built forms, any Cairenes would tell you, are mushrooming in and around Cairo. It is interesting to note that most of these global building types usually emerge first in North American suburbs. In Cairo, the urban sprawl, though in many ways haphazard, is indeed heading towards a similar suburban pattern. Of course, the state's decision to construct a ring road around the GCR and other major highways, like Empain's tram system at the turn of the twentieth century, opened up the surrounding desert for land speculation and lifestyle materialization.

In the fast rush for land that followed the deregulation of the media and the economy, the Bahgat Group managed to secure a 2,000 feddan stretch of prime real estate along the Cairo-Oases desert road, a few kilometres west of the great pyramids of Giza. Unlike the founders of Heliopolis, the creators of Dreamland did not need to excavate for an ancient Egyptian temple or the remains of a town in search for an authentic founding myth; the panorama of the pyramids was enough. In addition to the added value of this view, the cost of the project when completed is estimated to be US$ 2,640 million.

Like Empain, ninety years earlier, Bahgat searched for foreign architects to design his Dreamland. It is useful to mention that one striking feature that marks the architectural profession in the 1990s is the growing number of North American and European engineering and architectural firms practicing in Egypt: Bechtel, Sasaki, Michel Graves, SOM, are among many others. It was as if the last decade of the twentieth century was the marriage season for Egyptian and foreign architectural firms. Unlike Empain, however, Bahgat did not want to hire a foreign architect to design his Dreamland in the Arab style of architecture. Rather, he wanted a North American feel to it (Warson, 1999). Buildings were supposed to reflect the classical and the grandiose, American styles with pitched roofs and arches, all painted in light pastel colours.[27] Two North American offices were hired to develop the master plan and the architecture of the prospect project, Development Design Group Incorporated and FORREC, respectively. Both companies are global leaders in planning successful theme attractions and entertainment facilities. It is also interesting to note that both companies are participating in shaping not only Cairo's cityscape, but also that of other Arab cities; the former company is developing the Dubai Blue Coast resort, while the latter is building Ajman's Hypermarket and extending Dubai's Burjaman Mall.

Of course, the formula for designing Dreamland at the end of the twentieth century differed from that espoused for Heliopolis in the less globalized and more idealized decades of the early twentieth century. Whereas the creators of Heliopolis opted for the Garden City Movement as a source of inspiration, those of Dreamland wanted to dissemble 'the Orlando of the Middle East', referring to the Disney World development in Orlando, Florida (Matthews, 1999) (figure 7.3). The Disney Company is clearly a dominant player in the global entertainment market that, as urbanist Sharon Zukin puts it, 'not only represent an image of America that foreigners want to visit', but also signifies 'a way of life that others want to join' (Zukin, 1995, p. 50). No doubt, the creators of Dreamland recognized that no other industry worldwide has seen such constant expansion in recent history as the entertainment industry. In fact, not only the private sector, but also the state has been involved in expanding this industry. Just across the highway from Dreamland, the Ministry of Information is developing a theme park and a media production facility.

My point so far is that certain Western cultural icons, predominantly North American, are identified with a particular way of life, which came to be known as the American way. This way of life is taking over the rest of the world. Painted in shimmering pastels, these icons are mesmerizing people and their phantasms 'into one homogeneous global theme park, one McWorld tied together by communication, information, entertainment, and commerce' (Barber, 1996, p. 4). Dreamland is one intersection at the periphery of the expanding grid of McWorld, to use Barber's term.

From the start, Dreamland was intended to function as an independent residential city. Besides the various types of dwelling, its master plan reveals the intention of its makers to include many attractions and facilities: a theme park, shopping resort, polo pitch, office towers, hospital, two schools, two hotels, tennis and golf clubs, cinemas, theatre, amphitheatre, business park, and a human canon. The planning of Dreamland resembles a typical North American suburb with no clear centre or a downtown area. Rather the centre has been fragmented among a golf course, a theme park, and a shopping resort. In contrast to the boulevards and squares of Heliopolis, which organized the various elements of the city into clear geometrical forms, the designers of Dreamland opted for a hierarchy of twisting and curving roads with less formalistic, perhaps ambiguous, geometry. Another difference between the two developments, which,

Figure 7.3. Dreamland, the Disneyfication of the Middle East.

in my judgment, reflect a difference between capitalism then and now, is that Heliopolis was designed to accommodate spaces for production and consumption, Dreamland, on the other hand, is designed solely for consumption and fun. Moreover, where in Heliopolis foreign expatriates, tourists, and Levant and native government employees were rubbing shoulders yet all spatially segregated from the poor, in Dreamland, the new leisure community, which is predominantly Egyptian with a few expatriate Europeans, is rid of the poor altogether because of the price tag put on properties. But, these differences should not let us forget the similarities. For example, both developments depended heavily on their amusement parks and culture of fun to attract people to the development. The Heliopolis Luna Park was the first in Africa and the Middle East; in fact, it opened only five years after the original 'Trip to the Moon' opened at Coney Island. In promoting Dreamland, the company always projected images of Dreampark, stressing that it is the first and the largest theme park in Egypt and the Middle East (figure 7.4). In addition, both developments integrated the most advanced communication technology available in their respective times: the electric tram and the fibre optics.[28]

Reviewing the promotional materials for Dreamland and its master plan confirms the initial impression that the main intention of the project is to construct a place and time of fantasy, with a permanent air of festivity. Consider the names given to the different residential building types, for example. Hawaii, Bahamas Lite, and Caribbean are just but a few among many other designations that suggest the illusion of vitality and enjoyment through their semantic associations with famous North American destinations, known for their ambience of dissociated festivity. But the illusions are not limited to a particular geography or historical time. One is told in the brochures that in Dreamland one should expect to experience Vienna, Naples, or the historic district of Khan el-Khalili. In a kind of reverse tourism, the development is offering to bring the world and history to us. Of course, this effect of carnivalesque frivolity is particularly glaring in the project's centrepiece, Dreampark.

According to the chairman of FORREC, the Canadian firm had initially proposed ancient

Figure 7.4. The entrance to Dreampark, a hint of Universal Studios.

Egyptian themes for the park, but once again the developers of Dreamland wanted a North American style. In the end, the Park contained a small 'theme area more as an ancient wonders of the world attraction, and not specifically Egyptian'. The rest was a hodgepodge of 'Hollywood television programs and movies that most of the world watches and understands' (Warson, 1999). 'Themeing' is a strategy and a method of organizing space based on exploiting culture, and the Disney strategy has undoubtedly exerted influence on these designs. Like in the world of Disney, the tenets of these strategies, Zukin contends, combine visual culture, spatial control, and private management. And these are exactly the strategies, which Dreamland is trying to emulate. In fact, Zukin tells us that these are the strategies of the new symbolic economy, which I would like to add is supported by the neoliberal economic ideologies and manifested in theme parks and other cultural and entertainment, non-productive bases, which have been America's real economic growth engines (Zukin, 1995). This new economic reality seems to me the driving force behind the acceleration of leisure and entertainment avenues in Cairo during the last decade. Consider, for example, the growth of tourism and the urban restructuring of tourist sites. If Heliopolis represented the fledgling culture of entertainment in Egypt, Dreamland mirrors its latest fad.

My argument is precisely that what has been emerging out of the social and material production of Dreamland and similar late twentieth-century elite spaces in Cairo is a specific urbanity, which is linked to a particular form of economy, a neoliberal one. In this Disneyesque urban economy, experience has been controlled, organized, enframed, and linked to a fusion of consumerism, entertainment, tourism, and lifestyle (Hannigan, 1998). By enframed, I mean the tendency of this urbanity to wall off and privatize spaces. I want to make clear that a common thread runs between this new urbanity and the urbanity that guided the production of Heliopolis. My point is that from the early stages of capitalism in Egypt, entertainment and fantasy were interwoven with the production of elite spaces. Etymologically, the noun entertainment derives from the verb to entertain, or in Latin (*inter* + *tenere*), which meant to hold together. Spatially, the production of urban entertainment fantasies in this sense would be the holding together of delusive imaginations, generated through the use of particular signs and images. 'The holding together', I contend, is reached through a process of financial as well as spatial exclusions. This is how liberal economy and the urban oasis, or the neoliberal economy and the gated community are all linked together.

To shift the discussion slightly, I want to recapitulate part of the story and link it to one last point concerning the prevalence of this urbanity in all global cities. The 1990s deregulation of the economy and the media has been followed by a relative expansion of the symbolic economy and the service sector in Egypt. Once again, this economy works by developing places, within Cairo or at its outskirts, as urban oases for visual and cultural consumption: theme parks, media city, gated communities, water parks, drive-in theatres, multiplex movie theatres, and shopping malls. Although it is prevalent in many parts of the new satellite cities as well as tourist sites, this neo-urbanity, I want to make clear, is not dominating Cairo's urban scene. This is so, simply, because the symbolic economy is not dominating the city's economy. In fact, industry, agriculture, and a large informal economy still dominate the economy of Cairo and other major Egyptian cities. This being said, I want to suggest that projects like Dreamland represent Cairo's struggle to introduce this new urbanity, which is the city's membership card for joining the club of World Cities. It is an urbanity that evokes a lifestyle consisting of posh apartments, dream houses with swimming pools, golf courses, sleek restaurants, couturiers, and artistic culture, which is like a precondition for the global corporations to nestle in a city. Besides the Egyptian elite, the new urbanity caters for the lifestyle of global corporate executives. Zukin argues that these global employees are more attached to their lifestyles than to their corporations. That is why, in their struggles to attract the global corporations, I shall suggest, many cities across the Arab World, including Cairo, are adopting this form of urbanity.

Parting Thoughts

There is little evidence to suggest that Heliopolis was any more authentic than the Dreamland of today (figures 7.5 and 7.6). Both developments depended on the production of fantasy and entertainment for promoting their commodity, which is no less than space itself. I concede that to some people Heliopolis may seem more authentic because of the use of Arab architectural style. But let us not forget that both developments reflected the underlying economic and social processes of their respective eras. It may be the case that whereas foreign consortiums were the spearhead of the early adventures, a mix of Egyptian and Arab entrepreneurs are the progenitors of today's expansion inside and out across the fields and deserts surrounding the Greater Cairo Region. However, if anything, Heliopolis and Dreamland are deliberately formulaic and fabricated illusions of the Orient and the Occident, respectively. If the figure of the flâneur, the nineteenth-century idle stroller, fitted the image of a promenading European in Heliopolis, who combined urban voyeurism with a desire to consume oriental culture, today, the figure fits the image of a wandering Egyptian neo-bourgeois in

Figure 7.5. Roxy, the main shopping arcade in Heliopolis.

Figure 7.6. The main shopping plaza in Dreamland.

Dreamland, who combines a tourist gaze with a maximum impulse for consumption and a desire to indulge in the illusionary world of Disney. Whether fledgling or mature, the interaction of Egyptian and international capitalism has always been enticing the desires, which in turn stoke the construction of urban fantasies such as Heliopolis and Dreamland. This is, perhaps, why historical observers when examining Cairo's modern history do a double take, as if all this seems vaguely familiar: they are having a *déjà vu* experience.

NOTES

1. I am indebted to Samir Raafat for this observation.

2. Curiously, debates concerning foreign influences on local identity and traditions were as popular among politicians, journalists, and others a century ago as they are now (but under the rubric global-local debate).

3. Timothy Mitchell has suggested that in Egypt 'a conservative estimate is that within the last five years the area of its capital city has doubled'. In a public lecture given at the Association for Egyptian Architects, in autumn 2000, Dr. Tamer El-Khorazaty, a senior advisor to the Ministry of Housing and Urban Planning, suggested that in the past decade Cairo has quadrupled!

4. Emphasizing the continuation and the similarities between the two epochs will not be fully embraced by some researchers. For instance, the 'current break with modernity' advocates, who put more emphasis on securing the possibility of the uniqueness of our time: we live in a postmodern time. From their perspective, the relationship between globalization and Cairo would be rendered distinctly new because globalization signifies a fundamental rupture with past forms and sensibilities. Proponents of this position would argue that during the past few decades, both the developed and developing worlds have undergone the most significant transformations since the first decades of the twentieth century. These transformations are occurring at the economic, political, cultural, technological, and urban levels, and for which, nowadays, the term globalization serves as catchall to capture their elusive meanings and consequences. This position usually entails that the concept of globalization, in its contemporary usage, signifies a development in the transnational capitalism, hence, a rupture with modernity and modernization.

5. I want to suggest that this may partly explain why historical observers of Cairo have this *déjà vu* experience. There is a lingering threshold between, on the one hand, a drive towards a rupture with modernity and, on the other hand, a delay of the actual arrival of a truly postmodern time to paraphrase Lebbeus Woods.

6. From one point of view, the founding of 'Al-Qahira', Cairo, in 969 AD by the Fatimid dynasty was an immense expansion of the earlier settlement of Misr al-Fustat. The growth of the city during the Middle Ages

– from the camps of the Arab troops settling around the Byzantine fortress of Babylon in 640 AD to the founding of the Fatimid city of Al-Qahira in 969 AD – reflected the political shifts within the Islamic empire.

7. Timothy Mitchell argues that landholding referred to a system of ownership dispersed among the ruler, the ruler's local representatives, and the cultivators. Landholding, in this sense, referred to sharing land revenue and not to sharing land as an absolute object.

8. In this system a landlord, *multazim*, divided the agricultural land into categories, depending on the quality of the farmland. One category was left for peasants to farm for themselves with taxes collected at the end of the agricultural season. Another category, *Teen al-Wassiya*, was directly managed and run by the landlord with peasants working as waged or forced labourers. Other categories in this land division gave limited ownership for peasants.

9. Although in 1847 Mohamed Ali issued a decree allowing ownership of land granted to farmers, which gave farmers the right to mortgage land, exchange it, or lease it to others according to an official certified document, resorted to during land transfers or disputes, it was not until the Sai'idya Code that land was in truly freehold tenure.

10. It might be argued that the first profound and direct linkage between Egypt and Europe was during the short three-year French occupation of Cairo at the end of the eighteenth century. Later, Mohamed Ali brought many European nationals to work on different projects and sectors to regenerate Egypt.

11. Just before World War I, the export of cotton accounted for 93 per cent of total exports.

12. Because land speculation has always been a low barrier market, many adventurers entered to gain easy money. A typical story would begin with a group of financiers, usually a mix of Europeans and Levantine nationals, launching a joint stock land company, like the 'Giza and Rhoda Land Company' or the 'Building Lands of Egypt', for example. These companies would purchase a large tract, usually agriculture land close to the new tramlines, hold it vacant until the price went up and then, sell it for a much higher price. Needless to say, the assets of the companies in the stock market went up.

13. In around 1907, there were 94,925 thousand workers in the building construction sector, (see Beinin and Lockman, 1998, p. 38.)

14. The editor of the *Al-Ahram* daily newspaper in 1898 was very sarcastic about the rise of commercial advertisements, particularly how Westerners had developed all sorts of awkward methods to deploy them in our daily lives. One outlandish example, which he cited, was about an American who committed suicide by shooting himself with a 'Colt Pistol'. An arms' salesman took the opportunity and agreed with the family of the deceased to engrave on his tombstone the gun model and how effective it would be for those who like to emulate!' (The Art of Westerners in Advertisements, *Al-Ahram*, January 13, 1898).

15. There were many real estate companies. To the west and across the Nile in Giza, the Duqqi Land Company was developing the area close to Cairo University and along the Nile towards Imbaba; to the south, about 6 kilometres south of the city, there was the Delta Land Company developing the suburbia of Ma'adi; to the north-east, and parallel to the Ismailia canal, the Kubba Gardens Company was developing an area of about 110 feddans into a residential area; and, of course, there were other areas such as Garden City and Zamalek.

16. Shepheard Hotel, for instance, was burned and rebuilt several times, but during the 1880s it was built in an Italian style with covering of thick stucco.

17. In the north, namely, Shubra El-Khima, where the textile industries had existed since World War I, the government invested 5.3 per cent from its total industrial investment in metropolitan Cairo. In the south, near the Helwan area, the former winter retreat for royalty, it invested around 92 per cent of the total industrial investment, particularly in iron and steel factories.

18. Abu Zeid Rageh has suggested that the development of Cairo throughout history followed a distinct social pattern: the flight of the upper and middle classes to peripheral areas around the city, leaving the working and lower classes behind to occupy the centre. The centre continues to grow until it can no longer absorb the population to the point where it begins spilling out in the periphery; hence, the higher classes move to another outer ring and so on and so forth (Rageh, 1984, p. 178).

19. For instance, some of these measures appear in laws 43, 73, 136 of 1974, law 111 of 1975, and law 32 of 1977.

20. The history of this development of neoliberal capitalism has been examined at some length in the literature on political economy with reference to supranational powers given to the World Bank, IMF, WTO, and the like (Gill, 1995).

21. The notion of neoliberalism according to the

Thatcher-Regan doctrine combines aspects of liberal and conservative politics. While the liberal component promoted by these governments advocated liberalization, privatization, and internationalization of the economy, the conservative element centred on themes related to moral conservatism.

22. I use the word domesticated because I am fully convinced that the implementation of many of these economic agendas operate through a chain of personal as well as impersonal commands, which is never consistent with any abstract rule. Dr. Abd el-Men'im Said goes even further and argues that Egypt with the quasi implementation of the free market economy cannot be classified as a capitalist state (Said, 2003).

23. This is my translation of the title of Rushdi Said's book, which is in Arabic.

24. The housing supply by the state and the private sector cater for only a quarter of the actual demand.

25. We learn from the Bahgat Group's website that while completing his doctoral studies in the United States, Bahgat invented and manufactured an electronic device, which guided Moslems, regardless of their location worldwide, in determining the time and the direction of their prayers. According to this story, upon his return to Egypt in 1985, Bahgat capitalized on the success of his device and established a company specialized in electronics. Between 1985 and 1994, the company, International Electronics, focused on assembling Korean Television sets and other electrical appliances, and later using military factories to build its own products (Mitchell, 2002). It was not until 1995 that the Group expanded beyond electronics into real estate, tourism, and entertainment businesses. Today, the Group comprises more than forty companies working in many different fields: electronics, household appliances, furniture, medical, sports, entertainment, tourism, and real estate development.

26. For example, the state allowed the public usage of satellite dishes for the first time in the early 1990s. Later, it allowed the establishment of private channels.

27. According to one of my informants, the architectural styles of Dreamland were Bahgat's personal choices.

28. There are even similarities between the two developers. For example, both are industrialists who made fortunes before entering the real estate business. More over, they both built huge palaces for themselves in the middle of their developments.

REFERENCES

Abu-Lughod, Janet (1971) *Cairo: 1001 Years of the City Victorious*. Princeton: Princeton University Press.

Alam, M. Shahid (2003) Globalization and its discontents. *Ahram Weekly*, 12–18 June.

Amin, Samir (2003) The American ideology. *Ahram Weekly*, 15–21 May.

Amin, Samir (1997) *Capitalism in the Age of Globalization*. London: Zed Books.

Appadurai, Arjun (1994) The global cultural economy, in Williams, Patrick and Chrisman, Laura (eds.) *Colonial Discourse and Post-Colonial Theory*. New York: Columbia University Press.

Barber, Benjamin R. (1996) *Jihad vs McWorld*. New York: Ballantine.

Beinin, Joel and Lockman, Zachary (1998) *Workers on the Nile*. Cairo: American University in Cairo Press.

Bello, Walden (1996) Structural adjustment programs: success for whom? in Mander. J. and Goldsmith, E. (eds.) *The Case Against the Global Economy*. San Francisco: Sierra Club Books.

Chakravorty, Sanjoy (2000) From colonial city to globalizing city? The far-from-complete spatial transformation of Calcutta, in P. Marcuse and van Kempen, R. (eds) *Globalizing Cities, A new Spatial Order*. New York: Blackwell

Chambers, Iain (1990) *Border Dialogues: Journey in Postmodernity*. London: Routledge.

Chichian, Mohamed A. (1988) The effects of world capitalist economy on urbanization in Egypt, 1800–1970. *International Journal of Middle Eastern Studies*, **20**(1), pp. 23–43.

Chua, Amy (2002) *World on Fire: How Exporting Free Market Democracy Breeds Ethnic Hatred and Global Instability*. New York: Doubleday.

Crouchley, A. E. (1936) The Investment of Foreign Capital in Egyptian Companies and Public Debt. Technical Paper No. 12, Egyptian Ministry of Finance.

Economist Intelligence Unit (1996/97) *Egypt: Country Profile 1996/97*. London: The Economist, pp. 22–23.

El-Nagar, Said (1997) *Tagdid al-Nezam al-Siyasi wa al-Iqtisadi fi Misr*, Part I (in Arabic). Cairo: Dar al-Shorouk.

Fishman, Robert (1982) *Urban Utopias in the Twentieth Century*. Cambridge: The MIT Press.

Gauch, Sarah (2000) A widening class divide in Egypt. *Christian Science Monitor* (Online edition) http://

search.csmonitor.com/durable/2000/05/24/p1s3.htm.

Gill, Stephen (1995) The global panopticon? The neoliberal state, economic life and democratic surveillance. *Alternatives*, **20**, pp. 1–49.

Goldschmidt, Arthur (1988) *Modern Egypt: The Formation of a Nation-State*. Boulder: Westview Press.

Gran, Peter (1997) *Islamic Roots of Capitalism: Egypt 1760–1840*. Syracuse: University of Syracuse Press.

Handoussa, H. (1991) Crisis and challenges: prospects for the 1990's. in H, Handoussa and G. Potter (eds.) *Employment and Structural Adjustment: Egypt in the 1990s*. Cairo: American University in Cairo Press.

Hanna, Milad (1996) *Al-Iskan wa al-Siyassa* (in Arabic). Cairo: al-Hay'a al-Misriyya al-'Amma li-'l-Kitab.

Hannigan, John (1998) *Fantasy City*. London: Routledge.

Hassan, Fayza (2001) The setting of Sun City. *Ahram Weekly*, 541 (5–11 July), p. 16.

Howard, Ebenezer (1902) *Garden City of Tomorrow*. London: Swan Sonnenschein.

Ibrahim, Saad El-Din (1984) Al-Qahira, Hala Igtima'ya, in Serag El-Din, I., Fadil, R. and al-Sadik, S. (eds.) *Tahdiat al-Tawasou'a al-'Umrani: al-Kahirah*. Cairo: The Aga Khan Award for Architecture.

Ibrahim, Saad El-Din (1987) A sociological profile, in Nasr, S. and Hanf, T. (eds.) *Urban Crisis and Social Movements*. Beirut: The European-Arab Social Research Group.

IDMSC (Information and Decision Making Support Center) (1993) *Al-Mabani al-'ashwa'iya* (in Arabic). Cairo: Ministry of Local Government.

Ikram, Khaled (2001) Towards sustainable development, in *Emerging Egypt 2001*. London: Oxford Business Group.

Ilbert, Robert (1984) Misr al-Gadidh, in Serag El-Din, I., Fadil, R. and al-Sadik, S. (eds.) *Tahdiat al-Tawasou'a al-'Umrani: al-Kahirah*. Cairo: The Aga Khan Award for Architecture.

Ilbert, Robert (1981) *Heliopolis: genèse d'une ville*. Paris: Centre National de la Recherche Scientifique.

Issawi, Charles (1954) *Egypt at Mid-Century: An Economic Survey*. London: Oxford University Press.

Kellner, Douglas (1997) *Articulating the Global and the Local: Globalization and Cultural Studies*. Boulder: Westview Press.

King, Anthony (1990) *Urbanism, Colonialism, and the World Economy*. London: Routledge.

Knox, Paul (1995) Introduction, in Knox, Paul and Taylor, Peter (eds.) *World Cities in a World-System*. Cambridge: Cambridge University Press.

Lefebvre, Henri (1991) *The Production of Space* (translated by Donald Nicholson-Smith). Oxford: Blackwell.

Marsot, Afaf Lutfi Al-Sayyid (1985) *A Short History of Modern Egypt*. Cambridge: Cambridge University Press.

Matthews, Janet (1999) Construction special report, in *MEED Weekly Special Report*, (Online Report) http://web.nps.navy.mil/~relooney/Arabia_MEED38.htm.

Mekay, Emad (1999) Luxury housing market glut triggers regulatory calls. *Middle East Times*, May, p. 7.

Ministry of Housing (2002) *Mubarak wa al-'Umran*(in Arabic). Cairo: Wezarat al-Iskan wa al-Marafiq wa al-Mogtama'at al-'Umraniya.

Ministry of Planning (1997) *Al-Istrategiyya al-Qawmiyya lil-Tanmiya al-Iqtesadiyya wa al-Igtema'iyya fi Mostahal al-Qarn al-Wahid wa al-'Ashrin*. Vol.1 (in Arabic). Cairo: Ministry of Planning.

Mitchell, Timothy (1991) *Colonizing Egypt*. Berkeley: University of California Press.

Mitchell, Timothy (1999) Dreamland: the neoliberalism of your desires. *Middle East Report*, Spring, pp. 28–29.

Mitchell, Timothy (2002) *Rule of Experts*. Berkeley: University of California Press.

Muselhy, Fathy (1988) *Tatour al-'Assima al-Misriyya* (in Arabic). Cairo: Dar al-Madina al-Munawara.

Muselhy, Fathy (2002) *Al-'Umran al-'Ashwa'ifi Misr*, Vol. 1 (in Arabic). Cairo: Al-Maglis al-'A'ala lil-Thaqafa.

Overbeek, Henk and Van der Pijl, Kees (1993) Restructuring capital and restructuring hegemony, in Overbeek, Henk (ed.) *Restructuring Hegemony in the Global Political Economy: The Rise of Transnational Neo-Liberalism in the 1980s*. London: Routledge.

Owen, Roger (1970) *Abhath Al-Nadwa Al-Dawlyia Li-Tareekh Al-Qahera*. Vol. 1 (in Arabic). Cairo: Dar Al Kotoub.

Oxford Business Group (2001) *Emerging Egypt 2001*. London: Oxford Business Group.

Pevsner, Nikolaus (1976) *A History of Building Types*. London: Thames and Hudson.

Raafat, Samir (1996) When Maadi very nearly replaced Heliopolis. *Egyptian Mail*, November 16, p. 12.

Raafat, Samir (1997) Familiar ground: the nineteenth century privatization of Daira Sanieh doesn't seem that distant. *Business Monthly*, July, pp. 17–18.

Refaat, Mohamed (1989) Mashroa' al-Istefada min Tagrubat Misr al-Gadida fi Takhtit wa Tasmim al-Dhawahi al-umrania al-Gadida (in Arabic). Working Paper, Faculty of Urban Planning, Cairo University.

Rageh, A. Z. (1984) Al-Taghir fi al-Namat al-Iskani fi Madinat al-Kahirah, in Serag El-Din, I., Fadil, R. and al-Sadik, S. (eds.) *Tahdiat al-Tawasou'a al-'Umrani: al-Kahirah*. Cairo: The Aga Khan Award for Architecture.

Rodenbeck, Max (1998) *Cairo: The City Victorious*. New York: Vintage.

Said, Abd El-Men'im (2003) Misr Dawla Ra'ss Maliya! (in Arabic). *Al-Ahram*, July 7, p. 24.

Said, Edward (1979) *Orientlism*. New York: Vintage Books.

Said, Rushdi (1996) *Al-Hakika wa al-Wahm fi al-Waki' al-Misri* (in Arabic). Cairo: Dar al-Hilal.

Shahine, Gihan (1999) Reviving old splendour. *Al-Ahram Weekly*, 448, 23–29 September, p. 14.

Stewart, Dona (1999) Changing Cairo: the political economy of urban form. *International Journal of Urban and Regional Research*, **23**(1), pp. 128–129

Telkamp, G.J. (1978) Urban history and European expansion. A review of recent literature concerning colonial cities and a preliminary bibliography. *Intercontenenta 1*, Leiden: Center for the History of European Expansion, University of Leiden.

Warson, Albert (1999) Entertainment oasis coming to Sahara. *Shopping Centers Today* (Internet Search) http://www.icsc.org/srch/sct/current/sct9905/52.htm.

Woods, Lebbeus (1998) Inside the border, in Woods, Lebbeus and Rehfeld, Ekkehard (eds.) *Borderline*. Vienna: Springer Verlag.

Wright, Gwendolyn (1987) Tradition in the service of modernity: architecture and urbanization in French colonial policy, 1900–1930. *The Journal of Modern History*, **59**(2), pp. 291–316.

Yousry, Mahmoud, Abu-Zekry, Tarek and Yousry, Ahmed (1998) Cairo as a world city: the impact of Cairo's orientation towards globalization, in Lo, Fu-Chen and Yeung, Yue-man (eds.) *Globalization and the World of Large Cities*. New York: United Nations University Press.

Zordoq, Ahmed (1998) *Al-Tatour Al-Equtesady* (in Arabic). Cairo: Zagazig University.

Zukin, Sharon (1995) *The Cultures of Cities*. Oxford: Blackwell.

Chapter 8

Redrawing Boundaries: Dubai, an Emerging Global City

Yasser Elsheshtawy

The desert on the other side was, most of the time, flat, hard and desolate; nothing and no one came from its depth except rarely . . . The bedouin who had at first refused to go near the sea or take part in unloading cargoes from the small boats were soon won over. It seemed to them curious, arousing and somewhat risky, and before long they went closer to the sea. They did so hesitantly, in stages, with a sense of experimentation and secrecy . . . but their fear never left them, for 'water is treacherous, a swallower who never is satisfied' . . . Later on they waded into the shallow water. It was enticing, caressing their feet with its coolness and density, and with the passing of time they did not hesitate to bathe in the sea . . .

Abdelrahman Munif, *Cities of Salt*

The city-state of Dubai, a small emirate located on the Arabian Gulf, has been a crossroads for transient residents and travellers since its establishment as a small fishing village in the nineteenth century. Within the last two decades it has witnessed rapid urban growth due in part from the income generated from oil revenue, but mostly from various economic and industrial developments (figure 8.1). Its multi-cultural (hybrid) nature is evident from its unique population make up: a city of approximately 1 million residents, where locals form a minority (the number varies from 8 per cent to 20 per cent) while the majority are of Arab, Asian and Western nationalities. Given this unique population as well as its location at the tip of the Arabian peninsula, Dubai has become a border region in which one can detect a variety of 'conflicts': West/East, modernization/fundamentalism, Arab/Asian and so on. These conflicts are resolved spatially through a policy which on the face of it attempts to reconcile through co-existence, but a closer examination reveals an exclusionary direction through the development of clearly defined 'borders' i.e. zones or enclaves.

This is evident in modern projects such as the luxurious Burj Al-Arab (Tower of the Arabs) Hotel and the more recent Palm Islands development in which authorities have created an instant, easily recognizable image of a city which is on the verge of joining the global community (figures 8.2 and 8.3). Burj Al-Arab has, as this chapter will show, become a symbol of Dubai; in its imagery it evokes the sail, thus 'recognizing' Dubai's

Figure 8.1. The skyline of a newly emerging Dubai in the midst of a vast desertscape.

historical seafaring tradition. Located next to this tower is another mega project in which the city is embarking on a $1.5 billion development to create 'the largest man-made island' in the Arabian Gulf, a project which, it is claimed, is 'visible from outer space'. Shaped like a palm tree (in its site plan) the island is appropriately known as 'The Palm-Jumeirah' and will contain exclusive residences, hotels and shopping/entertainment facilities. The palm was chosen because it is 'one of Dubai's most enduring symbols of life and abundance' – another allusion to the past. Both of these projects are physically removed from the shore and are accessible via bridges, thus highlighting their exclusive nature. Their image – whether it is sails or palms – is directed towards the global media thus affirming what some have described as being 'the most globalized of all Arab societies'.[1]

This chapter, in critically assessing these projects, will first place them within the current globalization discourse with its emphasis on the notions of hybridity and the inevitable outcome of 'polarization'. This is followed by a brief historical background of the development of the city with a specific focus on the rapid growth witnessed within the last two decades, illustrating the dramatic urban transformations within the twentieth century while highlighting the city's multi-cultural nature. The current 'megaprojects' undertaken by the city are then discussed, and this discussion is followed by the two case studies. They are presented to illustrate the extent to which, firstly, they represent a continuation of modernization efforts in Dubai which began early in the twentieth century and, secondly, are a natural response to the current globalization discourse.

The Globalization/Hybridity Discourse

Many would argue that globalization has been part of the socio-political landscape in the Middle East since colonial times. Finding theoretical justification through disciplines such as 'Orientalism' – the study of Eastern cultures – which were meant to establish the superiority of the West, and an establishment of an 'other' which needs to be carefully studied, analysed and in

Figure 8.2. An advertisement for the Palm Island.

Figure 8.3. Burj Al-Arab Hotel.

turn 'domesticated' to fit with a 'universal' notion of appropriateness (Said, 1978). However, a theoretical shift has occurred recently whereby the cultural imperialism perspective and the ultimate hegemony of Western forms has been replaced by a 'cultural hybridization' paradigm.[2] What this view leaves out however is that there is an inevitable outcome of 'polarization', particularly in so-called 'world cities' which have come to symbolize the ultimate triumph of globalization. In such cities '. . . the view down below appears to be inhabited by a swirling mass of immigrant populations, competing for low wage jobs' and this is associated with a '. . . growing ethnic, racial and cultural heterogeneity, coupled with social and spatial polarization' (Oncu and Weyland, 1997, p. 1). The structure of these world cities has generated new forms of spatial and class polarization or what Friedman in his article 'The world city hypotheses' termed 'intersecting scape' (Friedman, 1986).

The case of cultural hybridization is made through the existence of various emerging cultural forms such as blended musical genres (the proliferation of the Algerian/North African *Rai* mixed with the Western form of contemporary rock, for example). A happy marriage between East and West is thus envisioned, effectively displaying a rather optimistic view of these global tendencies. Yet one might clearly argue against, and in effect refute, the theoretical shift paradigm: in essence the imperialist/hegemony perspective is alive and well. Furthermore, by changing terminologies the whole notion of a globalized world would become more palatable. Nowhere is this more evident than in various development projects in Middle Eastern cities. In this respect, efforts are made to respond to the global economy and to recreate the city through the 'tourist gaze'. The metropolitan experience is thus '. . . intricately involved in the global circulation of images' (Oncu and Weyland, 1997, p. 11). Traditional cities such as Cairo or Beirut effectively manufacture their heritage and culture – their 'image' – to attract investments, tourists etc. Increasingly, Arab oil cities – not burdened by history – are becoming models to be emulated. Their soaring glass towers, glitzy shopping malls, and various events are looked at as a 'true' sign of progress and modernization. This, however, poses questions pertaining to authenticity and socio-political equality. And – most importantly – are they truly 'hybrid'?

In these cities identities are constructed among expatriates, a continuation from the colonial period strengthened by continuing links with home countries – further enhanced by realities of the new 'global system' which enables one easily to maintain contact: easy travel, access to media. Within the United Arab Emirates, for example, each group maintains its own set of institutions. The existence of these groups has prompted many to claim that a hybrid culture was (is) being constructed in a process related to consumerism and identity (Elsheshtawy and Nagga, 2000; Kazim, 2000). For example, certain items from food or clothing find their way from one group to another. While this is the official 'image' that is being propagated – a happy merging of cultures – the reality does not sustain this view. In fact cultural groups maintain strict segregatory measures, by having their own set of institutions (schools etc.) which enable them to maintain (sustain) their cultural values. The public sphere – a setting allowing for a merging of these sub-cultures – is again segregated along economic lines (high-end vs. 'poor' shopping malls). Groups within these settings are clearly identifiable (Indian, Arab-Egyptian (-Lebanese, -Syrian, -Moroccan etc.)). Locals, clearly visible (by their traditional dress – *dishdasha* for men, *abaya* for women), form a distinct group . Intermingling is not encouraged – even within each cultural group, which would like to operate within its own constructed sphere.

While this should not be seen as an argument against hybridization it, nevertheless, is a suggestion that there is a negative outcome. Parts of the city are exclusively associated with 'elite' global elements – those that can afford the multi-million dollar (or euro, dirham etc.) price tag associated with these development/renewal efforts. The city is being created to respond to globalizing tendencies – tourism, large corporation headquarters, events (conferences, rock concerts, etc.) – thus increasingly catering to transient populations. No real effort is made to resolve social problems, address concerns of the lower class, or try to make the urban environment more 'liveable'.

Globalization has thus become another form of domination – primarily catering to those who have the necessary resources. While the imagery associated with the production of spatial forms is linked to cultural elements – an essential post-modern condition – thus suggesting hybridity, it ultimately caters to elite elements, thereby excluding ordinary citizens. Unfortunately, hybridity becomes a 'conspiratorial' actor by concealing its real intentions, cloaking itself in the mantle of tradition and (manufactured) heritage.

The Case of Dubai

During 2003 the International Monetary Fund

(IMF) and the World Bank held their meeting in Dubai, with more than 14,000 delegates descending on the city which had undertaken massive construction efforts, from a new convention centre to new roads, in preparation for the gathering. In many ways such an event marks a remarkable transformation from what used to be a small fishing village to a 'world city' in the making, rivalling its neighbours – Hong Kong and Singapore. This transformation which happened within only the last 30 years or so makes the emirate of Dubai a perfect setting for examining the tendencies discussed above.

Origins and Developments

The Gulf region has been an area of contention and struggle since the fifteenth century when the Portuguese, followed by the British, attempted to gain control over parts of what used to be then known as 'Historic Oman'. Motivation was primarily commercial, to control the vital spice trade an important commodity at that time due to its use in the preservation of food. The United Arab Emirates or UAE – as it is currently known – was initially part of Historic Oman. As an important trading centre it dates back to antiquity and Mesopotamian documents dating as far back as 2000 BC mention trading activities in 'Magan', as it was then known.

The Portuguese were the first Europeans (i.e. Western) powers to 'colonize' the Gulf in a period lasting from 1500 to 1750. Some have argued that the fall of Grenada in 1492 and the expulsion of Moslems from Spain followed by the Portuguese invasion of the Gulf could be seen as a continuation of the war against Islam carried out on Moslem lands (Smith, 1991). No doubt such ideological conflicts cannot be discounted but the area was also attractive as a centre of trade, located between European markets and those in Africa and Asia. While the Portuguese were emerging as a major power in the region other developments taking place were influential such as the fall of Constantinople in 1453 and the emergence of the Ottomans as another hegemonic entity. Meanwhile the Portuguese continued their expansion with a series of battles culminating in the capture of Hormuz, Persia in 1507 and subsequent domination of the entire Gulf area.[3]

A local population did exist in the area, however a series of tribal migrations occurred at the time which formed the nucleus for the current population's make-up. Tribes began to emerge from the desert and migrate to Historic Oman in the 1700s attracted by the promise of prosperity due to trade. Of significance are the *Qawasim* who came from Persia and the *Bani Yas* tribal grouping who came from the *Najd* area – in what is now Saudi Arabia. At that time the British showed interest in the area, and the *Qawasim* were perceived as a serious threat to their territorial ambitions. The influence of that tribe extended from Sharjah to Ajman, Um Al-Quwain and Ras Al-Khaimah (current Sheikhdoms – or Emirates). They also had a significant maritime fleet numbering close to 500 ships. The *Qawasim* formed an alliance with the newly formed *Wahabi* – or *Muwah'hidin* – movement in Central-arabia between 1795 and 1799.[4] The Ottomans and British came to see the *Muwah'hidi* expansion as a threat to their dominance. As a response to this alliance, and the *Qassimi* attempt to charge the British for navigational rights in the Gulf, it was decided to launch an attack against Ras Al-Khaimah in 1805. The *Qawasim* were ultimately defeated in 1819 thus establishing British colonial rule in Historic Oman which lasted until 1971 – although some have divided this period in two parts: 1820–1945 (end of World War II) and 1945–1971 (Kazim, 2000).

As opposed to downright physical occupation with all the responsibilities it entails, the British preferred to construct a series of treaties with a few select *Sheikhs* – tribal leaders. This was, of course, aided by the lack of any 'structured' state, or a unifying entity, a functioning civil, urban society as can be found in Egypt, for example, which would necessitate radical measures such as occupation. These treaties came to be crucial factors in the formation of the UAE and the emergence of various emirates such as Abu Dhabi and Dubai. One of the most significant is the 1820 treaty signed by the Sheikh of Abu Dhabi, to be followed by other Sheikhs including the Sheikh of Dubai which officially designated the British as the rulers of Historic Oman. This treaty extended to the Sheikhs' descendants as well. In 1922 another interesting treaty was signed specifying that in the case of the discovery of oil, concessions would not be extended to any non-British entity. This was prompted mainly by oil discoveries in Abadan, Iran in 1908 and in Trucial Oman during World War I. During British dominance in the area some resistance did exist which eventually culminated in independence, but these were short-lived and did not take the form of armed, organized resistance as could be found elsewhere in the Arab world.[5]

A number of significant consequences emerged from these treaties. First, Historic Oman was divided into two parts: Musqat & Oman and Trucial Oman. Trucial Oman was divided into separate Sheikhdoms, 'conferring' an identity on each – thus leading the way towards cultural, political and economic segmentation. Also, a social hierarchy was 'constructed' as a result of designating treaty signatories and their descendants as *de facto* rulers – with all the power and prestige that entailed. Through this shrewd policy power was conferred upon mercantile/trade families who signed these treaties thereby establishing hierarchical relationships – thus setting the basis for the current power base of *Shuyookh* (elder, rulers). Also, most importantly, these treaties separated the area from the rest of the Arab/Moslem world.[6]

The Foundation of Modern Dubai[7]

As noted, the *Bani Yas* were a competing tribal configuration with the *Qawasim*, their territory extending from Dubai to Qatar along the coast, and inland towards the Bureimi Oasis. The 1820 treaty segmented this tribe further by having Sheikh Tahnoon bin Shakhbout of Abu Dhabi and Sheikh Hazaa bin Za'al of Dubai (at that time 11 years old) each sign independently thus 'creating' the emirates of Abu Dhabi and Dubai. A significant development strengthening the independence of Dubai occurred in 1833 when – due to political turmoil in Abu Dhabi – the *Al-Bufalasah* section of *Bani Yas* migrated to Dubai eventually settling there. The *Al-Bufalasah* contingent in Dubai numbered almost 800 people, a figure which effectively doubled the reported population of the fishing village at the time. The Al-Maktoum family, current rulers of Dubai, were part of this 'wave of immigrants' who had decided to separate from Abu Dhabi. The mid eighteenth century saw expanded economic prosperity resulting in greater urbanization. While the bulk of the income was generated from fishing and pearling, Dubai's *souqs* were bustling with merchandise, in part as a result of the tolerance and openness of its rulers. Towards the end of the eighteenth century the population was estimated at around 3,000. Dubai's importance in the late 1800s and early nineteenth century was primarily due to the British bringing their goods via steamer and because it became Trucial Oman's main pearl exporter, after developments in neighbouring Iran.

In the 1870s political instability in Iran gave ports on the coast of Oman, among them Dubai, an opportunity to become the major trading centres of the area. In addition, due to tax increases in Persia commissioned by the Shah in 1902, Arab Sunni merchants migrated *en masse* to Dubai where they established a trading base. By 1925 the Persian taxes had become even more onerous and were seen to be permanent, and consequently the merchants who had been in temporary residence accepted the Dubai ruler's offer to settle in Dubai and brought their families over. These merchants of mixed origin were given an area of land immediately to the east of Al Fahidi fort on which to build themselves houses. The district became known as Bastakiyýa since many of the settlers came from the Bastak district, a part of the sub-province of Lār in the Fārs province.

Dubai, since its 'inception' attracted a variety of immigrants such as the Persian *Bastak* mentioned above. Other Persian and *Shi'i* groups existed as well. Bahraini pearl divers had their own special district – 'Firj al-Bahrainah'. Migrants came from Iraq to work as accountants and clerks for Dubai's merchants. Another round of migrants in the 1930s were *Baluchis* (from the Indian sub-continent) who worked as porters and in other low-level employment (Kazim, 2000). One could clearly argue that Dubai early last century contained several classes and ethnic groups, and these cut across lines of ethnicity and gender.

Britain's influence was strong because of its interest in maintaining ports of call in the Gulf for two reasons: first, as an alternative overland and maritime route between India and Britain and, second, to reinforce British presence to prevent Russian influence in Persia (Kazim, 2000). In 1904 steamers were calling at Dubai on a weekly basis, encouraged by the abolition of a 5 per cent customs duty thereby declaring Dubai a 'free port'. The city also became a distribution centre for goods to a number of cities along the Gulf as well as the interior of the country and the Sultanate of Oman. The city itself was divided by the creek in two parts: Deira and Bar Dubai. Deira, to the north, was considered the mercantile sector, Bar Dubai to the south was the government centre housing such buildings as the customs house and the residence of the Sheikh. Ferries connected the two parts.

During the 1947–1971 period, following World War II, Dubai continued to maintain itself as a mercantile city. However it still remained more or less a British protectorate, a status enhanced by a series of political measures and development projects. For example, numerous contracts were extended exclusively to British firms; in 1947 the British Bank of the Middle East was established, in addition to the construction of Al-Maktoum hospital, a joint project with the British, in 1953. At that time Dubai replaced Sharjah as the seat of the British political agent.[8]

Under Sheikh Rashed, a series of development projects – all carried out with British assistance – greatly helped in strengthening the city's position as a centre of trade. These included the electrification of Dubai in 1961, a year which also saw the dredging of the creek for which funds were borrowed from Kuwait to pay for the British firm of Halcrow & Partners. In this project the port was deepened through dredging and the building of breakwaters. Finance for many of these projects was arranged in anticipation of oil revenue. Oil in commercial quantities was discovered in 1966. British companies to which all these contracts were extended needed cheap labour which was readily available from India and Pakistan. Many came from other parts such as Iran, Europe and Arab countries. In 1968 immigrant labour constituted 50 per cent of Dubai's population. These workers had considerable cultural impact but they had no political power within Dubai's civil society – for example, they were not permitted to form unions. The influx of these 'foreigners' strengthened the identity of locals in relation to these immigrants. Also 'Dubayyans' perceived themselves to be superior to northern sheikhdoms due to their economic privilege – despite the fact that they all lack substantive quantities of oil.

In spite of the presence of diverse cultures the city enjoyed relative stability – a fact paradoxically enhanced by fragmentation. Each societal segment became too self-contained to be able to unite with others and in turn challenge the *status quo*. Furthermore, tribal divisions were less significant in Dubai than in Abu Dhabi, for example, which is where most of the tribal groupings presided.

Forms of Resistance and Anti-Colonialism

Attempts at reform and in turn removal of colonial power did occur. In 1930 armed conflict occurred between Sheikh Saeed Maktoum and the *majlis* – a grouping of merchants acting as advisor to the Sheikh and led by the trader Sheikh Maani – which eventually led to a British intervention, the installation of a new *majlis*, and Sheikh Maani going into exile. Also termed the 'Reform Movement' they were pearl merchants reacting to the declining income from pearl revenue. Other mercantile groupings refused to join however. This event prompted some to declare that '. . . behind Sheikh Saeed was British colonial power' (Kazim, 2000, p. 203). Interestingly, the origins of this movement can be traced to a variety of factors such as the opening of the Suez Canal, development of steam shipping and a postal system, all of which led to the spread of ideas through the media prevalent at the time such as newspapers which advocated nationalist movements and the construction of an 'Arab' identity in relation to the colonial powers. Another relevant event pertains to the construction of the airport. The British, encountering resistance in Ras Al-Khaimah, set their sights on Dubai – at the end of 1931. Many of the Sheikh's relatives as well as merchants wanted to prevent Dubai from becoming the outright centre for British hegemony in the Gulf. However, a small airport accommodating flights between Alexandria and India was finally built in 1937, thereby establishing a British presence in the city.

The period following World War II saw a weakening of British hegemony and the emergence of the United States as a replacement. This was of course linked to a rising global anti-colonial movement in the period between 1945 and 1965 which manifested itself in forced withdrawal from many Arab countries. This was in contrast to what happened in the Gulf, however, which witnessed an early British voluntary withdrawal for a number of reasons. The more peaceful transition here ensured that oil concessions would remain intact and that British firms would be hired for maintenance and expansion of infrastructural installations. A stable political structure in place ensured the continuity of ruling entities and prevented the emergence of an armed insurrection.

However, in spite of all this, the region did witness anti-colonial activities which, while not reaching a stage of armed insurrection, were nevertheless critical in establishing – or constructing – a 'UAE' identity among others. The first stage of this struggle was, of course, the short-lived reform movement in the 1930s which was not explicitly anti-colonial, its focus being mainly on governmental reform. The second stage, however, made British withdrawal one of its key demands

– deriving inspiration from Egypt's Nasser. A third – armed stage – as happened in Oman and Aden – did not occur mainly because oil revenues led to massive development projects resulting in prosperity and the beginnings of consumerism. Furthermore, the lack of the development of an indigenous working class contributed to this. Workers, brought by the British retained identification with their home countries.

On 2 December 1971 what had been known as 'Trucial Oman' became the United Arab Emirates. There were many unifying factors which smoothly paved the way for this transition: the same series of treaties imposed on each emirate; pearl fishery; tribal interconnections; and, of course, being subject to the control of the British government of India. The occupation of the islands of Greater and Lesser *Tunbs* by Iran – one day prior to the announcement of Independence (!) – helped in strengthening the newly emerging UAE identity.

The Establishment of Dubai as a Global Centre

The narrative so far has discussed the formation of the UAE, and Dubai, by examining various historical, social and political factors. It has shown that Dubai developed an economy based on international commerce because it has been playing the role of a commercial *entrepôt* since the late 1800s. Attesting to the ingenuity of the local population, they have been able to overcome the decline of the pearling industry due to the manufacture of artificial pearls in Japan and following the great Depression. They began a re-export business using the tax-free zone and the Dubai Creek. Due to the dredging of the creek, and the discovery of oil noted earlier, the city witnessed an unprecedented expansion and construction boom. Oil revenues were used to spur infrastructure development such as schools, hospitals, roads and a modern telecommunications network. A new port and terminal building were built at Dubai International Airport. A runway extension that could accommodate any type of aircraft was implemented. The largest man-made harbour in the world was constructed at Jebel Ali, and a free zone was created around the port which was an ingenious response to changing market conditions. In contrast to transnational corporations operating within Oman toward the end of the colonial period, the ones operating in the 1980s and 1990s were more flexible, provided more rapid service, worked with more resources and changed more quickly according to market conditions, both within the UAE and globally. Nowhere is this more evident than in Jebel Ali – Dubai's free zone – where these corporations could operate freely and without restrictions.

Managers and stockholders are the main beneficiaries of Jebel Ali free zone but there are other segments (strata) as well – these include low-wage workers and service sectors: accountants, bank personnel, market analysts who link the transnational corporations' Dubai headquarters to the 'world'. One of the main attractions, however, is the availability of low-wage, non-unionized workers. In addition, companies have been able to avoid some of the tax and environmental regulations which may exist in their home countries. Also, there are no numerical restrictions on expatriate workers nor any wage parameters. The last factor was particularly significant in the 1980s and 1990s when emphasis was on manufacturing which was labour intensive.[9]

Thus the city set the stage for becoming a major global centre for trade and commerce at the end of the twentieth century by providing the necessary infrastructure for global corporations. All of this had major spatial implications on the urban growth of the city and on its architecture which will be discussed in the following section.

Contemporary Dubai: Urban Expansion and Rapid Urbanization

The massive and rapid expansion of Dubai is the most tangible manifestation of urbanization that has transformed the city from a small fishing village to a major city. This 'spatial expansion' is manifested in size, appearance, and urban morphology (figure 8.4). Three factors contributed to this growth: economic affluence, technological advancements in transportation and public utilities, and government land-use policy (AlShafieei, 1997). Before discussing these factors it is of interest to note some numbers pertaining to the status of Dubai at the turn of the twentieth century. At that time the population of Dubai numbering 10,000 was centred in three areas: Deira which consisted of 1600 houses and 350 shops whose population was primarily Arabs, Persians and Baluchis; Shindagha which had 250 houses and only Arab residents; and Dubai which had only 200 houses and 50 shops dominated by Persian and Indian merchants. There are no detailed data as to the character of these houses except that they were built of clay and coral fragments. In 1956 the first house using concrete blocks was built (Gabriel, 1987).

The first phase of urban development extended from 1900 to 1955. Dubai's urban growth in this period reflected a pattern of slow and limited physical expansion due to constrained economic activities and marginal growth in population. The entire population of the city was confined to three small enclaves located at the mouth of the Creek. Both the Creek and the Arabian Gulf were considered as the main source of income for the local population who were predominantly involved in fishing and pearl diving. Until 1955 the urban area did not exceed 320 hectare (3.2 km^2) and the land use was predominantly

Figure 8.4. An aerial view of Dubai. Sheikh Zayed Road appears in the foreground; in the background is the traditional city centre.

residential with extremely limited commercial space. The vast majority of the local inhabitants were living in extended families in *'Barasti'* houses (houses built of palm fronds). Residential quarters were built in clusters to serve the purpose of privacy and collective tribal security. The internal parts of each enclave were connected by narrow walkways intended for pedestrians and domestic animals that were used for transport of goods and drinking water. The only source of drinking water was four public wells near the enclaves and water was carried directly by the families (Heards-Bey, 1982). The second phase of Dubai's urban development can be characterized as compact growth, extending from 1955 to 1970. Developments followed a master plan prepared by British architect John Harris in 1960. It called for the provision of a road system; zoning of the town into areas marked for industry, commerce and public buildings; areas for new residential quarters and the creation of a new town centre. These rather modest goals were in line with the emirate's limited financial resources (oil had not yet been discovered in sufficient quantities). In 1971, due to the city's expansion and the availability of economic resources, a new master plan by Harris was introduced. The plan called for the construction of a tunnel running beneath the creek connecting Bur Dubai and Deira (Shindagha Tunnel) and the construction of two bridges (Maktoum and Garhoud); in addition the building of Port Rashid was also envisioned. A large area extending toward Jebel Ali was designated for residential use (now known as Jumeirah) and a further large area was reserved for industrial use; to the city's south additional areas were dedicated to health, education and leisure/recreation uses (Gabriel, 1987).

Dubai is now undergoing what some have called 'planned suburban growth', a period of rapid expansion which began in 1970 (AlShafieei, 1997). Of particular note is the city's growth corridor along Sheikh Zayed Road towards Jebel Ali. Dubbed as the 'new Dubai', this area is emerging as the new commercial and financial centre of the city. Numerous projects are constructed along this stretch of highway and the skyline of the city is changing as a result. These rapid developments have emerged both as a result of increased resources and an attempt to provide alternative sources for revenue.

In the 1980s and early 1990s, the city took a strategic decision to emerge as a major international-quality tourism destination. As a result investments in tourism infrastructure have paid off over the years. Recent statistics suggest that visitors numbered three million, greatly exceeding the one million population. According to some observers it is '. . . perhaps the world's fastest-growing global city and intercontinental hub' but it is also a '. . . non-Arab society planted in the heart of Arabia' (Hirst, 2001).

It is this last point, frequently cited, which is of great interest, particularly if one examines the population make-up. The city is composed of an immigrant population which, proportionate to the natives, is the highest in the world. In the UAE as a whole, with a total population of 2.5 million, nationals are outnumbered seven to one by mainly non-Arab aliens: 1.2 million Indians, 600,000 Pakistanis, 100,000 Iranians and contributions from dozens of other nations, including 50,000 Britons. In Dubai, nationals fall to a mere 8–20 per cent out of a million-plus. This has prompted some to term it not a melting pot but a '. . . polyglot polity in the making' (Hirst, 2001). The society however is open-minded, business is encouraged and recently freehold ownership has been opened for expatriates.[10] This atmosphere of openness has encouraged investment and tourism. Furthermore, additional sources of wealth have contributed to this rapid urbanization process: expansion of traditional commerce, increased regional business representation, in addition to tourism.

The Creation of a 'Global' Image

One observer recently commenting on the developments taking place in Dubai noted: 'Mapmakers in Dubai must be constantly frustrated. No sooner have they finished their latest in-depth representation of the city when another major project is announced to send them scuttling back to the drawing board. The map of Dubai is constantly being rewritten' (Smalley, 2002). These projects are based on a recognition that oil resources will eventually dry up. Therefore, the city has embarked on a multi-billion dollar drive to establish itself as the Gulf's hub for leisure, services and commerce. This involved increasingly 'bold' architectural projects: Palm Islands, Burj Al-Arab Hotel, Dubai Marina, and the twin Emirates Towers. Other projects are geared to the global economical infrastructure of the media and internet: Dubai Media City and Dubai Internet City (DIC) and Dubai Inter-national Financial Center (DIFC). Other cities are in the making: Healthcare City, Textile City and International City (initially called Chinatown).

Described by some officials as 'megaprojects', they distinguish themselves by using superlatives to ascertain their uniqueness. Furthermore, these projects are seen as a continuation of major developments in the city which contributed to its rapid urbanization – the dredging of the Creek thus facilitating commerce, and the creation of the largest man-made artificial port, Jebel Ali, a free-zone and industrial area located north of Dubai. The following is a brief account of some of the more visible of these projects, which will provide a contextual framework for understanding the importance and significance of the two case studies.

The Megaprojects

We already live in a town of cities: Internet City, Healthcare City, Media City, Festival City. Is the Sheikh Zayed road soon to be renamed Skyscraper city? Perhaps eventually each activity will have its own city, and we will find ourselves visiting Hamburger City for takeaways, Gritty City for car washes, Ditty City for CD's and music . . . we do wonder though why the most important thing of all – knowledge – only warrants a village and not a city.

Editorial, *scene* magazine, Dubai, March 2003

In discussing the major projects in Dubai – in essence the spatial characteristics of its recent expansion – it will be useful to group them according to certain themes. In other words some projects are geared towards the communication industry with its emphasis on Information Technology and the 'New' Media; another group falls within the domain of traditional mixed-use real estate developments; some projects provide a setting for an emerging financial industry, and of course the ever growing hospitality/leisure industry. The retail sector, with its accompanying shopping malls is the most visible among this group. Cultural projects such as museums, cultural centres and the like are, interestingly, lacking. Many of these projects are located along a stretch of highway linking Dubai to Abu Dhabi known as Sheikh Zayed Road, a spine which will form the centre of the 'New Dubai.'

IT/Media. Dubai has emerged as a leading centre of Information Technology in the region, a fact which is accentuated annually through its GITEX exhibition (an IT trade show). However, a more proactive response was through what has become known as 'Internet City' which was one of the first 'megaprojects' to be built. Announced by Dubai's crown prince Sheikh Mohamed Maktoum in 1999 it was completed in less than a year. According to official statements 'The mission of

Dubai Internet City is to create an infrastructure, environment and attitude that will enable Information and Communications Technology (ICT) enterprises to operate locally, regionally and globally, from Dubai, with significant competitive advantage'. Within the short time span in which it has been operating it has become home for some major companies who have set up regional offices; for example, Microsoft, Oracle, Canon and many others.[11]

Comprising a series of office buildings overlooking an artificial lake and lush gardens, the city is located adjacent to Sheikh Zayed Highway, a newly formed corridor along which the city is currently expanding. Entry is through a gate which is designed to reflect a traditional wind tower and a series of screens which contain a set of 'Islamic' motifs, and leads to a series of glass buildings such as one may find in any high-tech park in Malaysia or Silicon Valley for example. The entire project is protected by a fence although entry is free to anyone (provided they have a car) (figure 8.5).

Located nearby is Media City, a similar arrangement to Internet City, although in this case related to the Media Industry. This city is also comprised of office blocks within an artificial landscape. It houses studios, newsrooms and has become a regional centre for many media companies such as Reuters, CNN, MBC and others.[12] The anonymity of the office blocks in either of these two cities, only distinguished by the logo of their inhabitants, highlights the fact that they operate primarily on a global level, in some way disconnected from the surrounding reality (figure 8.6)

Mixed-Use Real Estate Communities. The city has its share of these large-scale developments which have become a defining feature of global cities everywhere. Of the more ambitious is one known as Festival City – due not only to its size but also its location. This waterfront 'city-within-a-city' spans 3.5 kilometres along the shores of the Creek and is set to transform, according to its developers, shopping, living, leisure and working lifestyles in Dubai. Designed as a centre-place destination, Dubai Festival City, a privately funded project of the Al-Futtaim Group, will offer a 'world-class' and 'unmatched' mix of entertainment, dining, shopping, a marina, hotels and residences, as well as an 18-hole recreational

Figure 8.5. Entrance to Dubai Internet City.

Figure 8.6. Entrance to Media City.

golf course, and sports and leisure facilities. It will have fifteen communities, each with subzones, which will be developed in line with the Emirate's market demand. The masterplan also includes a 50-storey 'Promontory Tower', which according to officials, will become another Dubai landmark because of its 'iconic design'.[13]

The project's location along the Dubai Creek is a 'vivid reminder of the long sea-going tradition of this Arabian port, representing not only Dubai's geographical centre, but also its birthplace, from which the modern Emirate emerged'. Thus the city is '. . . respectful of the past, yet designed for the future'.[14] It is located in an area known in Dubai as Deira, the traditional mercantile centre and is close to City Center, a Mega-mall development by the Al-Futtaim group. Also, the 'city' is relatively close to the centre of Dubai and has not been built in the 'new Dubai' to the north which is the site for many new projects. In this way the development is attempting to create an aura of accessibility which it also claims in advertising the project. Contextual references are made by noting that the development is based on two natural 'contextual features' – the desert and the sea – which are 'interwoven' throughout the plan to 'deliver powerful human experiences'.[15]

Further away, close to Dubai Internet City, the world's 'largest master planned waterfront development of its kind', Dubai Marina is EMAAR's, a governmental real estate company, large-scale project where construction is well advanced on building the region's first 'intelligent city' that will eventually house 35,000 people in luxurious residential apartment towers and villas in a 'spectacular' waterfront location. This man-made marina is strategically located on the Arabian Gulf in close proximity to leading five-star hotels, beach clubs, international restaurants and prestigious landmarks such as the Dubai Internet City and the Dubai Media City. It is approximately 3.5 km long with 10 km marine frontage. Spanning an area of some 2,800 square metres, the project comprises 10 districts, each of which will be developed as a distinct community with a mix of hotels and apartments.

According to company officials 'the most advanced information technology systems will be available in every home as standard'. Already over 1500 apartments have been sold in the towers that have been offered for sale. The first phase of the Dubai Marina project, Marina 1, comprising six towers housing 1100 residential units and equipped with the most advanced technology, was completely sold out, soon after it was put on sale. The first three towers have been named after precious stones, namely 'Murjan' consisting of 40 storeys, 'Al Mass' with 31 storeys, and 'Fairooz' which will have 23 storeys. The other 3 derive their names from Arab scents, 'Mesk' with 40 storeys, 'Yass' with 27 storeys, and 'Anbar' which will have 19 storeys. To identify these towers the top of each has been designed to reflect – again – an Islamic pattern. According to the architects of the project, initial designs have been kept intentionally simple yet owners have argued for a more 'forceful expression' of imagery – hence the 'islamisized' top.[16] Additionally there will be 64 villas and up to 40 retail outlets.[17]

Financial Centres. An initiative announced in 2002 was the creation of a financial centre. This entails a multi-billion dollar real estate development on 450,000 square metres of desert, featuring a 50-storey headquarters and up to 14 other skyscrapers.[18] Intended as a financial market for a region comprising the GCC, the Indian subcontinent, Caspian States, the Levant and east Africa, according to an official declaration the centre is '. . . perfectly situated between the financial centers of Europe and Asia, Dubai International Financial Center (DIFC) is a bridge between the east and the west serving as a gateway for the flow of capital to and from the region thus strengthening the regional links to international capital markets'.[19] Described by some officials as a 'missing piece in the puzzle', the centre is intended to complement other financial markets such as Frankfurt and Singapore.[20] The project has not been without its critics however, with some noting that it is just another example of 'desert parceling'.[21] Overall, however, the project is seen by officials as a way for Dubai to establish its presence on the global map by claiming to serve '. . . a market of 1.5 billion people'; it also is seen as a continuation of other initiatives such as Internet City, Media City, etc.[22]

Construction began with The Gate an imposing office complex, modelled on the Parisian Arc office building in La Defense, Paris. Similar to its Parisian counterpart the opening in the building is meant to establish an axis with a city landmark, in this case the World Trade Center, one of Dubai's 'oldest' modern landmarks built in the 1970s. The gate will lead to a gigantic open plaza which will be lined with a series of office buildings, media screens and other modern urban amenities – more or less establishing a new city centre.

Office/Hotel Complex. The city's skyline, long dominated by the World Trade Center building, has undergone a dramatic transformation, which has occurred within the last 10 years. A defining landmark are the twin buildings of the Emirates Towers, an office/hotel complex. At 51 storeys and a height of 350 m for its office tower it is the tallest in the Middle East and Europe (figure 8.7). Other projects are being constructed nearby which will include Burj Dubai, the tallest building in the world, claimed to be a 'city within a city'. The project's developer is EMAAR properties and the design has been awarded to Chicago-based Skidmore, Owings and Merrill (SOM). Official statements describe the design as being inspired by '...the cultural and historical influences of the GCC [Gulf Cooperative Council]' and this is achieved in the tower's base where the '. . . geometry reflects the six petal desert flower of the region'.[23] It is hoped that the building will surpass the tallest buildings in the world such as the Petronas Towers in Malaysia and

Figure 8.7. The Emirates Towers.

the Shanghai World Financial Center (it is still not known whether it will also exceed the height of the Taipei Tower in Taiwan considered on its completion to be the tallest but whose exact height is being kept a secret). However, this building, according to EMAAR's chairman Alabbar, will '. . . create a new architectural landmark'.

Retail Sector. Retail occupies a large share of Dubai's built environment and seems to be the single most dominant development taking place. Activities such as the Dubai Shopping Festival, Summer Surprises etc. are in fact geared towards these centres. The most visible among the malls is al-Futtaim's Deira City Center which has undergone a massive expansion within the last few years, thereby occupying a large chunk of Deira. A hotel has recently opened which is linked directly to the centre. Malls such as these are constantly being developed, expanded and improved. One of the more innovative which has recently opened is the Mercato mall which in its architecture represents a medieval European cityscape, complete with signs of deteriorating paint. To understand the importance of these malls, in particular as they resonate with the city's historical mercantile tradition, the current expansion of the Burjuman mall is of particular interest. A Dh 1.4 billion (0.4 billion US$) development, it will, according to officials, 'take its place in UAE history as one of the world's most breathtaking retail developments'. History and tradition are 'essential' elements of this development which, by 'maintaining the core elements of Arabic heritage and tradition while introducing the latest architectural designs and concepts', provides a 'beautiful look of modernity without losing its connection to the past'. The project is supposed to 'usher in a completely new shopping culture' by being 'the one place for discerning shoppers of luxury brands'. Thus it will enhance 'Dubai's image as an international shopping and leisure destination'.[24]

The foregoing offers only a glimpse of the projects currently under construction. Hardly a day passes without an announcement of a new 'megaproject' promising its uniqueness in the world; for example, 'Dubai Pearl', a 'landmark downtown concept' which will include the Gulf's first opera house in addition to a mixture of residences, retail and leisure facilities, is to be located next to Internet City and the Palm.[25] Another project aims at creating an underwater hotel.[26] Yet all of these activities have not been without their critics. There is concern that a 'real estate bubble maybe inflating'. While it is argued that demand for office space is high some have noted that 'this demand is generated by firms migrating from lower-quality buildings and many of these older buildings in less fashionable parts of town are increasingly sparsely occupied'.[27] Thus, while most of these developments are geared towards 'New Dubai', the traditional city is in danger of being neglected – a characteristic of many urban global centres.

Evoking Symbols: The Sail and the Palm

The sheer scale and enormity of the monumental Palm development embodies the spirit of Dubai as we forge a bright new future into the 21st century. The Palms are by far the largest man-made islands in the world and form complete, virtually self-contained communities, richly indulging all earthly desires for comfort, convenience and overwhelming beauty. The Palms are a feat of ingenuity, imagination, and engineering which is sure to become one of the modern wonders of the world.

From the Palms' development promotional pamphlet

It is easy to exhaust superlatives when describing Burj Al-Arab, a hotel that has already become an international landmark. Burj Al-Arab's sail-shaped structure, soaring 321 metres above the Arabian Gulf, has become as much a point of reference in Dubai as the Eiffel Tower is in Paris.

From Burj Al-Arab Hotel's website

While the projects discussed previously are impressive in terms of scale and size they are nevertheless not seen as national landmarks or symbols. In other words, these megaprojects are perceived primarily through their economic impact and the revenue they may generate. However, within this list two projects stand out: Burj Al-Arab (Tower of the Arabs) and the Jumeirah Palm Island, a hotel and a real estate venture. These two projects are being used as symbols to represent the new Dubai – while aiming at economic revenue as well, their semantic value, particularly in their choice of architectural symbols will offer some interesting insights within the context of this chapter. It will be shown how, through architectural devices, images are created which are directed at a global audience and this is done intentionally to further highlight the fact that Dubai is a global city. Issues of heritage, identity even authenticity become meaningless expressions in such a context.

Tower of the Arabs

Burj Al-Arab, or Tower of the Arabs, is a 56 storey (28 double height floors) 7 star hotel containing 202 suites, three restaurants, a conference centre and health spa. The hotel is built 290 metres out to sea on a man-made landscaped island (figure 8.8). The accommodation wings enclose two sides of a triangular atrium that runs the full height of the building. Its height is 321 metres making it the tallest hotel in the world and the fifteenth tallest building. The hotel contains a helipad at a height of 212 metres above the sea, a 60 metre mast at the top and a cantilevered restaurant 27 metres long and 200 metres above sea level. The total size is 111,600 square metres. Access is via a heavily controlled bridge. Only hotel guests and those who can afford the expensive entrance fee

Figure 8.8. Burj Al-Arab extending into the sea.

are allowed entry. The helipad is for those arriving from the Dubai Airport – as an alternative guests are whisked away in Bentleys.

The hotel was designed by British firm W.S. Atkins and its chief architect Tom Wright. They claim that the hotel is the 'ultimate landmark of Dubai and indeed the region'. Furthermore, the project had to be 'Arabic, extravagant and super-luxurious and – most importantly – become a symbol for Dubai'. Emulating such world city symbols as the Eiffel tower, Sydney Opera House etc. the architect looked at:

> . . . historical-arabian themes like wind-towers before realizing the potential of the seafaring theme. But instead of picking on the time-honoured shape of the dhow sail, as many expected, he fine-tuned the design to incorporate the profile of a J-Class Racing Yacht, itself a symbol of opulence and technology. Eventually he had a design which would be easily recognizable worldwide as Dubai's Burj Al-Arab – Tower of the Arab – a balance of east and west.[28]

In describing the design process the architect notes that his main aim was to find an 'iconic form'. His search eventually led him to discover that an easily, instantly recognizable form would be the proper approach.[29] Regarding the issue of culture, the architect, after noting that initially he did not know the location of Dubai argues that its people were forward looking aiming at the future, hence his decision to move away from the 'limiting' historical context.[30]

Stepping inside the hotel atrium – claimed to be the world's tallest – one is confronted with gold coated pillars, extravagant fountains, and brightly coloured furniture pieces – in fact the atmosphere of 'A Thousand and One Nights'. The sheer extravagance and scale is unprecedented – perhaps to justify the steep price tag (approximately \$2000–\$5000 per night).[31] This opulence is justified by being cast as a form of traditional Arab hospitality.[32]

Planning for the hotel began in 1993 and it was officially completed in 2000. Since then it has become the 'official' symbol of Dubai. Its image is used extensively in the advertising industry, as well as replacing the 'old' symbol of Dubai – the Fahidi fort and Sheikh Saeed house (one of the first built structures in the city) – on car licence plates. The project has not been without its critics. While some have noted its extravagance others have pointed out that it displays a schism in its dealing with local culture, by trying to 'have it both ways' – suggesting 'deep cultural roots' even though the real point is to show the city's sophistication and modernity – and that the project primarily aims to convince people that Dubai '. . . has more to offer than sand and distant views of oilfields'.[33] Also, in official publications the surrounding context is never clearly shown. The hotel appears isolated in the midst of a wide expanse of water – somehow hiding the inescapable and harsh reality of the surrounding desertscape. Edward McBride, expressing the official architectural reaction to this project, noted that the hotel is symptomatic of an era where 'the medium is the message' although unlike in Las Vegas, where one has roulette wheels, magic shows or night clubs, here the 'spectacle' is the structure itself (McBride, 2000). He further noted the projects exclusiveness: '. . . the management has tried to heighten the hotel's allure by preventing people from visiting it' (p. 118). He describes it as a 'ritzy haven from Arabia's sandstorm' and that the hotel has become a symbol of 'the triumph of money over practicality' by 'elevating style over substance' (p. 125). The project was thus designed from the top down by starting from a desired image and moving to its physical manifestation. Such a reaction is typical from professional architectural circles yet it belies the real intention behind the creation of these icons

– namely the construction of identity which will be discussed in the conclusion.

An Island shaped like a Palm Tree

Located near the Burj Al-Arab Hotel is the Jumeirah Palm Island, a palm tree shaped resort island on land reclaimed from the sea (figures 8.9 and 8.10). The official numbers and statistics pertaining to the project are quite impressive: it will add 120 km of sandy beaches and will include 2000 villas, up to 40 luxury hotels, shopping complexes, cinemas and a marine park. The island will be built in the shape of 17 huge fronds surrounded by 12 kilometres of protective barrier reefs, extending 5 kilometres into the sea south of Dubai city. They will be accessible by 300-metre bridges from the mainland or boat to two marinas, while the main causeway will also have a monorail system.

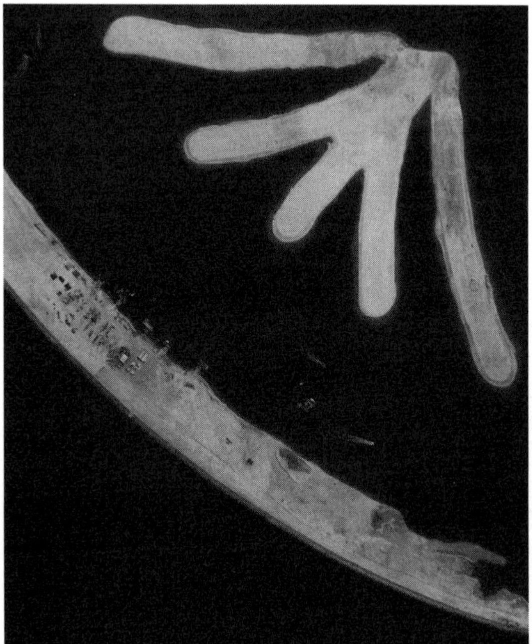

Figure 8.9. The outer crescent of the palm.

The island is claimed to be the biggest man-made island, and '. . . like the great wall of China it will be visible from the moon'. The palm was the brainchild of Dubai's crown prince Sheikh Mohamed bin Rashid al-Maktoum, essentially responding to the expected increase in tourists. To cater for them additional shoreline was needed and the island – measuring 6 km by 5.5 km will give Dubai 120 km of new, sandy coastline. According to the Palm Development Company the project will '. . . promote Dubai worldwide, it will encourage tourism, it will encourage business people'.[34] The date palm design was chosen because it is '. . . one of Dubai's most enduring symbols of life and abundance'.[35]

Interestingly the project was conceptualized by a New Zealand architect based in Australia – Warren Pickering. He was asked in 1995 to create a design for a proposed island. His proposal for an island shaped like a palm was eventually adopted. During three years of planning the project 'involved a total of 42 consulting firms and more than 50 studies, all especially commissioned to include key aspects such as marine ecology,

Figure 8.10. Aerial View of the Palm Island.

human population and the business development of the islands' (Belbin, 2002, p. 29). While the need for additional shoreline is understandable, the various implications of this project are not quite clear, specifically ecological ones – even if claims to the contrary are made. In fact according to one official, these are all details that will eventually be worked out, what matters is the 'right attitude'.[36]

According to official publications, to reflect the 'global outlook' of Dubai, the development company is offering 32 architectural styles for its residences ranging from Islamic to Mediterranean to modern. These are shown to potential buyers in the form of external perspective images. Following the sale, owners may change the interior. According to the architect, the villas represent a new approach in residential architecture in Dubai. Rather than being enclosed on all four sides by a wall, as one may find elsewhere, only three walls are built and the fourth side will face the sea so that 'people . . . enjoy the benefits of living by the sea' – hardly a revolutionary concept but nevertheless offered as an incentive for buying.

In addition to these residential units 50 'boutique' hotels will be constructed as well. These will offer different themes and will be built on the 'crescent' an 11 kilometre long crescent shaped breakwater protecting the residential fronds from the seaward side of the island. All hotels on the crescent will be low-rise not exceeding three storeys.

The architectural style for some of the elements of this project is described as 'Arab-Eclectic'. According to US-based architect Larry Ziebarth (whose clients include Disney, Universal Studios and Sea World) this style offers a 'sensitive blend of Islamic tradition with new architectural elements'. His company HHCP Design International has been asked to inscribe this style on some of the developments taking place on the palm. Further, describing this 'style', he notes that: 'We have sought to create a building style that is deeply rooted and borrows from regional Islamic traditions, but with a strong dash of fantasy – which will become synonymous with the entire project' (Belbin, 2002, p. 30). The result, as can be seen in the Palm Sales Center located on the shore and facing the island, is a 'Disneyesque' creation – not surprising given the architect's other clientele (figure 8.11). It is interesting that all these descriptions are primarily directed at laymen – what is in essence described as Arab eclectic is nothing but a reminder of postmodern excesses of the 1970s and 1980s which are finding

Figure 8.11. The 'Arab-Eclectic' style of the Palm sales office.

a re-affirmation in Dubai. That description, however, has been adopted by officials and is being used extensively in the media. Also, the literal and figurative are used not only in the site plan of the islands but extends to the design of some of the major buildings such as the Palm Centers – which will include a transport interchange for boats and a retail setting – where the design is 'inspired by the Nymphea or Water Lily and resembles a large flower on a reflecting pool'.[37]

The project is marketed as a means for individuals to realize a dream of owning property.[38] Yet at a price tag of $500,000 plus, that dream may not be in everyone's hands. Aware of such criticism, that the island caters for the super-rich or will become a 'rich man's retirement colony' an official notes: 'I don't think it is only for retired people and it is not only for the rich either. The rates are very reasonable and many will be able to afford them'.[39] According to recent sales records all villas have been sold and there is a large waiting list of potential buyers.[40] Initially offered on a 99-year lease basis, authorities decided to sell the units as freehold. Thus, buyers will own their properties outright and, to encourage sales further, they are also given an unlimited residency visa. The last fact is of particular importance since Dubai consists mainly of a transient population, hence having the option of unlimited stay is particularly attractive – conditional of course on job stability and having the ability to pay the high mortgage rates.[41]

The project is managed by Dubai Palm Developers and dredging work is undertaken by a Dutch firm (Van Oord ACZ). Some of the statistics associated with this construction are quite impressive. Palm Jumeirah will be built using 95 million cubic metres of sand and rock. Rocks for the foundations are, according to official records, taken from quarries around the UAE but also, amazingly, from Iran. Special ships, barges, are being used to dredge sand from the sea and – using Satellite GPS technology – sand is deposited to create the crescent shaped fronds. The project is moving according to schedule and it is expected to be completed in 2004 (figure 8.12).

As a footnote to the above, further developments are being created. The Palm Jebel Ali, a similar island as Jumeirah built further north, is set to become an entertainment destination

Figure 8.12. The outer crescent of the Palm – a ground level view.

and offers some unique housing typologies. These include houses located on the water. The arrangement of these houses will reflect a poem by Sheikh Maktoum.[42] Apparently the sale of these 'water homes' has been a huge success.[43] Also, the Palm Company is building another island called 'The World' comprised of a series of small islands whose overall shape will be that of a global map. Each island may thus represent a country or a continent. In advertising the project, the company highlights the exclusive nature of this development in that it is only accessible by boat. The chairman of the developer, Nakheel Properties, notes that these islands are for an investor who would 'wish to own his own island' and that these islands will 'consolidate Dubai's credentials in the super exclusive list of global destination spots'.[44] The company is also responsible for non-water/island related projects such as Chinatown (now International City), a mixed-use project, with the aim of developing the 'emirate's increasing trade and tourism interest with China'.[45]

Fragmentation and Mythical Environments

While these projects are admirable in their sheer scope and ambition, they tend to highlight the downside of globalization. One might argue that they represent a form of hybrid development – at a purely formal level, but also in terms of adding value to the city and the country – they nevertheless emphasize 'polarization' and 'exclusiveness', an unfortunate by-product.

The two projects have a number of commonalities. First is the use of superlatives to describe them: whether it is the 'tallest' hotel or the 'largest' man-made island.[46] In using such attributes the distinctiveness of these developments is ascertained, thus setting them apart. Their exclusive nature is further highlighted by locating them offshore, and controlling access through a variety of means: security guards, entrance fees, and by their sheer extravagance, prohibiting (or inhibiting) ordinary folk from entering (figure 8.13). In that same vein, the projects are located outside the relatively dense urban context of Dubai – thus playing a large role in the development of 'New Dubai' which will supposedly replace Singapore or Hong Kong as the financial and leisure hub of Asia.[47] In terms of imagery, the hotel and the islands are using easily recognizable symbols – the sail and the palm. While the sail can be seen from a general view, the palm is essentially a planning device meant to be seen from the air – not an everyday occurrence. Thus, the purpose is for this 'image' to be used as a marketing ploy, but also to become a symbol for Dubai. In the case of the hotel, the sail shaped image has replaced the traditional Sheikh Saeed house on all Dubai licence plates (figure 8.14). The palm aerial view is used in advertising and media campaigns for a variety of companies and financial institutions. While it is claimed that these 'symbols' are inspired by local and traditional elements, they acquire a more universal character: the sail, or the palm, are associated with many places throughout the world and are not necessarily exclusively linked to Dubai or the UAE.

The use of direct symbols such as can be found on various roundabouts and in other public spaces (the 'coffee pot' or '*dal'la*' appearing periodically in towns throughout the UAE) reflect a desire to connect with the past and to construct an identity *vis-à-vis* a predominantly migrant population.[48] The Palm Island or Burj Al-Arab could be understood or comprehended as a continuation of such thinking, albeit on a gigantic, i.e. global, scale. In such a way the stage is set for Dubai as a global entity. Simple, humble developments will

Figure 8.13. The gated entrance to the Burj Al-Arab.

Figure 8.14. A car licence plate on which the Burj Al-Arab is placed.

not achieve such an effect. Projects will have to be big, unprecedented, and constructed quickly. Issues of scale (large) and aesthetism/symbolism (easily understood and communicated quickly, no time for abstract reflections) are becoming paramount!

Furthermore the perpetuation of 'mythical environments' is another feature shared by these developments – which occurs at a variety of levels and scales. The Burj Al-Arab reflects an aura of 'A Thousand and One Nights' opulence and decadence; also a simulation of a submarine ride to reach a seafood restaurant is one of its dominant features. However, the Palm in particular has taken this concept to interesting levels. The residences being sold are distinguished by created stylistic forms (Mediterranean, Islamic, Asian, etc.) thus offering buyers a 'created' environments catering to their desire – the formal language for establishing stylistic links is, of course, purely personal (effectively mythical); furthermore, the propagation of these mythical environments is taken below water in developing coral reef landscapes based on different regions in the world (the Red Sea for example); visitors may thus move from one underground environment to another without having to leave Dubai.

This water theme is of particular significance. Arabs have, traditionally, shunned the sea in favour of the endless horizons of the desertscape – yet they have always been fascinated by water. In many respects both the Palm and the Burj are attempts to connect with the sea and to move away from the desert thus becoming 'modern' and 'advanced' by overcoming the perceived image of a nomadic people confined to the desert – similar to Harran Munif's fictional settlement in *Cities of Salt*, a seaside city constructed by the 'Americans' to house the Bedouins, relocated from their original inland settlements. The sea is initially viewed with suspicion yet with time becomes a source of fascination. Here, however, the settlement is initiated by the colonial power and Munif's nomads have to go to incredible lengths to overcome their fear of building and living near water!

Conclusion: Redrawing Boundaries

In trying to place these developments within the larger context of globalization, questions of 'urban identity' ultimately must be addressed. Yet the unique situation of Dubai, unburdened by history and heritage, led to a *tabula rasa* approach whereby the city is constantly being re-invented. In the course of this progressive 'redrawing of boundaries' there has been an apparent '. . . evacuation of the national' (Oncu and Weyland, 1997, p. 10), and fragmented developments are geared towards a global audience. The city is thus being re-created through the gaze of the tourist, the look of multi-national corporations, and the stare of real estate development companies. In such a context it becomes quite natural for developments to become isolated and fragmented islands (figure 8.15). McBride argued that the overall effect of these projects is one of a 'film set' and not a city. While there are plenty of extras enacting scenes within rows of tall buildings – outside the frame is a 'trackless desert'. He notes that 'Dubai has built itself the body of a city without the soul' (p. 125). However, and this is perhaps one of the main objectives of these projects, for many residents in Dubai they are a source of pride and a symbol of national identity. In fact they are used as tools to construct a national identity since they 'reinforce the local identity among UAE citizens in relation to others . . . and link [them] to a certain version of (*sic*) past' but this identity-creation also serves pragmatic functions by '. . . attracting tourists and investors to the UAE thereby linking the UAE to (*sic*) new globalism' (Kazim 2000, pp. 428–429). But in doing so divisions are created by reinforcing the identity of a privileged segment (multinational corporations and their beneficiaries) versus an 'other' (low-wage immigrant workers). Given Dubai's unique social structure, in which its fragmentation is in fact a source of its stability, such divisions may not prove as problematic – but the problem lies in the city's (underestimated) influence.[49]

These images are being emulated by other (more traditional) cities in the Arab world. Suzanne Kassab in her analysis of post-war Beirut noted that the reconstruction of its old business centre follows a tabula rasa approach which relies on 'fragmentation' in part inspired by Arab oil cities (Kassab, 1997). Cairo is following a similar course as Farha Ghannam argued, whereby the global discourse has led to a 'displacement of the local'. She notes, in discussing the displacement of local residents in Cairo from a poor neighbourhood, that:

> . . . many . . . families experienced the force of the global in their displacement from their 'locality' in the centre of the city. Their houses were demolished to be replaced by buildings and facilities that cater to upper class Egyptians, international tourists and the transnational community . . . relocation [is] utilized as part of the state efforts to 'modernize' Cairo and its people. (Ghannam, 1997, p. 120)

Recent developments in Cairo also suggest that Dubai is continuing to exert a strong influence. In 2002 Al-Futtaim group opened a City Center, complete with a Carrefour supermarket, on the edge of Cairo – near to what is being called New Cairo. The incongruous nature of such a development, particularly as one moves through Cairo and observes the advertisement for the centre, is striking.

Thus, while hybridity suggests that questions

Figure 8.15. The Dubai skyline with Bur Dubai – the traditional city – in the foreground.

of identity are finding some form of re-affirmation, by examining developments in newly created cities – those located at the East/West divide – one discovers that globalization may in fact lead to an erasure and a reconstruction of identity and a recreation of the city to respond to universalizing tendencies. The city of Dubai does provide a perfect example for such a course, yet in many ways it also offers a model for other, more traditional cities. It should however, in its strive to become a global centre, uncover 'the real Dubai' which does exist along the shores of the Creek and its marketplaces. This 'identity' needs to be strengthened, not in a superficial manner, but in a way which would make the city liveable for all.

NOTES

1. Adulkhaleq Abdulla, UAE University as quoted in *The New York Times*, 14 October 2001, 'Sweeping a tale under Gulf land's pretty rug'.

2. For a discussion of this issue see Oncu and Weyland (1997). See also, AlSayyad (2001) and Jacobs (2001)

3. It is interesting to note that Aden, Yemen remained in Arab/Moslem hands throughout the entire Portuguese period.

4. Named after its founder Mohamed Ibn Abd Al-Wahab (1703–1791) who advocated a 'return' to pure Islamic values and abolition of 'inventions'.

5. An interesting event known as the 'Hyacinth incident' occurred in 1910. The Sheikh of Dubai at that time – *Butti Ibn Suhail*, refused British demands for the installation of a British political officer, telegraph station and post office. In turn he was accused of hording 'illegal' arms, which led to an armed incursion into Dubai supported by the British warship, the Hyacinth, causing the death of 37 locals. Interestingly the charge of 'illegal' weapons was never proven (Kazim, 2000, p. 143).

6. Different identities were constructed among the various Trucial Oman sheikhdoms evidenced in the use of currency. During the period 1820 to 1945 all of Trucial Oman used the Indian Rupee because they were administered by the government of British India. After Indian independence in 1947 all sheikhdoms adopted a special Gulf Rupee. In the 1960s separate currencies were adopted. Abu Dhabi used the Bahrain Dinar whereas Dubai used the Qatari Riyal (Kazim, 2000, p. 224).

7. There is no consensus on how Dubai or *Dibai*, as it was widely known until relatively recently, was named. Of two theories widely espoused by historians, the possible favourite is that Dubai is drawn from *Dab*, the Arabic name for a spiny tailed lizard which was common in the desert of the Lower Gulf, although little known today because they avoid urban areas. Others believe that Dubai is a derivative from an Arabic word meaning 'land devastated by locusts'.

8. As further evidence of British dominance they argued in the 1960s that given the various treaties signed with 'Trucial Oman' they had the 'right' to be their sole representative in the UN.

9. The port is currently undergoing massive expansion which will increase capacity from the present 4 million 20-foot containers to 21.8 million to be finished by 2020. In 'Dh4.2b Jebel Ali Port expansion plan unveiled', *Gulf News*, 27 January, 2003.

10. At the time of writing this chapter there were still some questions pertaining to the legality of this. The UAE federal law does not allow ownership by foreigners (see '*Dubai – but do be careful*' FT.com, *The Financial Times*, 16 August 2003). A 'one country two systems' formula – similar to Hong Kong – has been proposed ('Hong Kong model solution to Dubai's land ownership issue' *Khaleej Times*, 29 July, 2003).

11. The city is viewed as a completely independent entity to highlight further its global status and connectedness to a wider regional and international network. It is advertised as follows: 'It is the Middle East's biggest IT infrastructure, built inside a free trade zone, and has the largest commercial Internet Protocol Telephony system in the world. Dubai Internet City is a strategic base for companies targeting emerging markets in a vast region extending from the Middle East to the Indian subcontinent, and Africa to the CIS countries, covering 1.6 billion people with GDP $1.1 trillion. In line with Dubai's liberal economic policies and regulations, Dubai Internet City offers foreign companies 100 per cent tax-free ownership, 100 per cent repatriation of capital and profits, no currency restrictions, easy registration and licensing, stringent cyber regulations, protection of intellectual property'.

12. Overall there are 23 TV and radio companies which are helped by a 'favourable business environment for

broadcasters to develop and flourish'. In 'Dubai Media City gets Dh1b investment', *Gulf News*, 23 June, 2003.

13. Lee Tabler, CEO Dubai Festival City Development in 'A Tale of Two cities,' *www.stars.com*.

14. Omar Al-Futtaim, CEO of Al-Futtaim Group; *ibid*.

15. Lee Tabler; *ibid*.

16. Babak Eslahjou of CORE (Core Architects Inc) Toronto, personal communication.

17. Other projects included in the EMAAR portfolio are Arabian Ranches which has an equestrian theme, and the gated communities of Emirates Hills, which has a golfing concept. In 'Building dreams into reality', *Gulf News*, Supplement, 11 May, 2003.

18. The project is located behind the Emirates Towers complex, and Shaikh Zayed Road.

19. From the official DIFC website – www.difc.ae

20. 'The basic premise is that there is a piece missing in the puzzle. There isn't a financial center between Frankfurt and Singapore', said Dubai International Financial Center (DIFC) spokesman Geoff Rapp', *International Business*, October, 2002.

Another official noted that DIFC 'will be a center like London, Frankfurt, Singapore, Hong Kong and Tokyo', stressing it would not be competitive but complementary to those markets. 'We are not creating another offshore unit, but a center at the same high financial level as established international financial markets', he said, adding that the market was expected to be in business by the end of 2002. *International Business*, March, 2002.

21. One Dubai-based fund manager, however, described the initiative as 'another example of the government parceling up the desert'. 'Any initiative has to be welcomed as a step in the right direction, but it seems at the moment that it's another real estate venture whose only added value will be an alleviation of the country's red tape situation.' *International Business*, October, 2002.

22. 'After having successfully established Dubai Internet City, Dubai Media City and having launched *Tejari* and other technology enabled initiatives such as e-government, the launch of DIFC was the next logical step and an indispensable second pillar for making Dubai and the UAE a major hub in the global "knowledge economy"' (*Gulf News*, 17 February, 2002, p. 30). The project has already attracted international attention as witnessed in a recent *International Herald Tribune* article which notes: 'Dubai is set to reach maturity as a global financial and business center by establishing the region's first money, capital and reinsurance markets. These new markets, brought together under the Dubai International Financial Center (DIFC), will add a new dimension to an emirate that has already built a reputation as the region's trading, tourism, sports and information-technology capital. Blue-chip companies and financial institutions from all over the world are now established in Dubai . . . These companies are attracted by Dubai's pro-business philosophy, cultural diversity and first-class infrastructure, which includes the information technology demanded by international business.' (*International Herald Tribune*, 20 May, 2002).

23. 'Emaar opens registration for Burj Dubai; names Chicago firm for design consultancy.' *Gulf News*, 22 May, 2003. Also see 'Tall Order', *Identity*, August/September 2003 – a real estate promotional publication based in Dubai.

24. 'Burjuman to usher in new era; size and scope of the project have been enhanced by the finest features in architecture and technology.' *Gulf News*, 23 June, 2003.

25. 'Work on Dubai Pearl to start soon.' *Khaleej Times*, 24 August, 2003.

26. 'Dubai plans latest epic destination, with 220 suites under the sea.' *The Guardian*, 14 August, 2003.

27. 'Dubai real estate sector to see over supply – EIU report.' *Khaleej Times*, 14 August, 2003.

28. In *'Atkins wins world's best new building award.'* (www.atkingsglobal.com). Also the architect describes the process as follows: 'We were seeking an image that was evocative, recognisable, and expressed these desires. The Arabian Gulf had always been a major focus for the city and now with the fast emerging tourist trade reinforcing the maritime association it seemed reasonable to design the tower hotel in the form of a giant sail. Not the shape of the historic dhow but the shape of a modern high tech sail, the type of sail that would be found on a yacht in St. Tropez. The type of sail that has always evoked a sense of luxury, excitement, sophistication and adventure. To complete the design, we would build the sail on an island offshore creating a building that we hoped would instantly be recognisable the world over'.

29. It is of interest to quote the architect's description of, and in turn the significance placed upon, the icon: 'The tower was to become an icon for the developing Dubai, as Sydney has its opera house and Paris the Eiffel Tower, so Dubai was to have a similar landmark. It was soon realised that to create a building that would become iconic was the most difficult aspect of the conceptual design. A list was drawn up of instantly

recognisable buildings that had become synonymous with a city or country. The prerequisite of the list was if the form of the building could be drawn and recognised within 10 seconds, pictorial style. The list was surprisingly small and most of the buildings were historical. Try it for yourself, you will be lucky to find 10 buildings in the world that fall within the category. You will find that each of the buildings has one thing in common, an economy and uniqueness of form'. (Chew and Schmitt, 2000, p. 123)

30. The architects described this as follows: 'It didn't take long to realise that the people of Dubai were enjoying their newly found freedom and wanted to be seen as a fast developing forward looking "first world" culture. This being the case we decided to look at the future aspirations of the people for our ideas rather than basing them on the limited historical cultural context'. (Chew and Schmitt, 2000, p. 124)

31. A recent article describes the hotel as follows: 'the Burj Al-Arab, where Russian gangsters with suitcases of cash and Saudi sheiks vie to occupy palatial suites with gold-plated taps, gold leaf banisters and screaming colour schemes of crimson, citrus and carrot. The exterior of the Burj Al-Arab is shaped like a full-blown spinnaker, and its outside walls are covered with a milky Teflon-coated fabric as protection from the searing sun . . . the hotel consists of 202 duplex suites. The two royal suites, which rent for $8,000 a night, are occupied only occasionally. The basement seafood restaurant, enveloped by walls of an aquarium and reached by an elevator designed as a faux submarine, is not always full either. But who cares? The hotel, with its revolving bed in the royal suite, is more talking point than hostelry. Nelson Mandela is said to have been dismayed by the opulence. But Bill Clinton, twice a guest, talks up Dubai as an exemplar of the possible marriage between Islam and the modern world'. (Khan, 2002)

32. An official promotional publication notes the following: 'In designing the Burj Al-Arab one had to have an understanding of the location; its culture, people, climate and future. Equally important is Arab hospitality, which is recognizably second to none. In Arabic tradition, each visitor to one's home should be received and welcomed as if he or she was a manifestation of the Prophet. In its recent times the older generations can still recall the nomadic Bedouin tribes that roamed the stark desert, graciously offering food and shelter to passing caravans. Here at the Burj Al-Arab this tradition is being followed by its seven star service . . .'. (Chew and Schmitt, 2000, p. 7)

33. In D. Sudjic 'Is that room service? Where am I?'. *The Observer*, 20 August, 2000.

34. A Palm official notes that 'this is definitely the most exciting time ever in Dubai's real estate history; for the first time expatriates can buy their own property and really feel a part of this boom . . . the increased demand for property has been a major boost to the economy. It is creating thousands of new jobs in the construction sector as well as other professions such as architecture and interior design, and, of course, the hospitality industry'. (Belbin, 2002).

35. 'Biggest man-made isles rise from the Gulf'. *The Times*, 5 January, 2002.

36. 'Our problem was that Dubai had run out of beach', Mr. Gergawi said. 'We were brainstorming one day. Someone said, "Why don't we make an island". But just one blob in the sea was insufficient. The plan for a series of finger islands arranged in the shape of palm trees evolved as the best way to maximise the amount of beach front', he said . . . Environmental concerns? Mr. Gergawi said they had been taken into account. Even if every detail of how so many people and structures would exist on slender slivers of sand is not quite clear, it is the attitude that matters.' Khan, A. (2002). 'Dubai Financial, Transportation Hub'. *The Nation, Online Edition* (www.nation.com).

37. In 'RMJM to design Palm Centers; firm is selected from short-list of 11 architects'. *Gulf News*, 10 May, 2003.

38. According to officials UAE nationals account for 75 per cent of buyers while 5 per cent constitutes other Gulf nationals. Of the remaining, 15 per cent comprises long-time UAE expatriate residents and 5 per cent is from overseas. Other estimates place nationals at a much lower percentage.

39. 'Paradise regained'. *Khaleej Times*, 11 May, 2002.

40. In a 2002 newspaper article a sales agent notes: 'There is so much demand from buyers, it is hard to meet the requirement with one island. Palm office currently has a waiting list of over 800 firm orders. With almost 90 per cent of the units put on sale already sold, they believe the 1,100 units on the nine fronds will be sold out in the next 10 days', said Zubair S. Mostafa, Marketing Manager of Al Khayat Real Estate. 'Buyers comprise mostly Arab nationals, Europeans and people from the [Indian] subcontinent who apparently prefer the Palm island to other beachfront properties in Cyprus and Greece. Some 70 per cent of the buyers are UAE nationals'. 'Palm project records biggest-ever booking wait list'. *Khaleej Times*, 21 May, 2002.

41. According to official statistics buyers comprise Arab nationals, Europeans and people from the subcontinent (India). No specific numbers exist, although officials note that 70 per cent are nationals – an unlikely number given the population make-up of the city. Other sources indicate that the majority of buyers are expatriates – mostly GCC nationals. *Khaleej Times*, 21 May, 2002.

42. A newspaper article describes this as follows: 'The new residences will be laid out in a pattern resembling a verse from an Arabic poem penned by General Sheikh Mohammed bin Rashid al-Maktoum . . . which translates into: "Take wisdom from the wise people; [*sic*] Not everyone who rides is a jockey".' M. Nair, 'Palm introduces residences on water concept in Jebel Ali'. *Gulf News*, 6 May, 2003.

43. This is reflected in numerous newspaper articles such as 'Investors lap-up Palm Water homes; over 40 pc sold on first day'. *Gulf News*, 9 May, 2003; also 'Water homes draw good response'. *Gulf News*, 7 May, 2003.

44. Sultan Ahmed bin Sulayem as quoted in 'There is a market for individual islands'. *Gulf News*, 12 May, 2003.

45. The project is located in the *Al-Aweer* area of Dubai. 'Nakheel Properties to develop Chinatown'. *Gulf News*, 12 May, 2003.

46. A distinct Dubai characteristic. A 2001 newspaper account notes that 'It rejoices in superlatives. The *Guinness Book of Records* is on permanent stand-by to record them. Normally they apply to the biggest and best of everything that "Maktum Inc", as some call it, can boast. It may be the tallest and most luxurious hotel in the world, like the phantasmagoric Burj Al-Arab, recently completed on a man-made island in the Gulf, the world's largest man-made port at the Jebel Ali free zone, or the largest-ever wedding cake'. Hirst, 2001.

47. A 2001 local newspaper article cites a British academic study which claims that '. . . the rise of UAE's profile in the international arena is challenging Singapore's stature as a premier "global city"'. 'Rise of UAE profile worries Singaporeans'. *Gulf News*, 1 February, 2002.

48. The persistence of the national dress – a free flowing robe – called *dishdasha*, as well as the predominance of certain forms of dialect as well as the discouragement of intermarriages (specifically between female locals and foreign males) are symptomatic of this phenomena.

49. The last few years have witnessed a plethora of newspaper articles, attempting to describe and analyse this newly emerging city. For example, the well known German newsmagazine *Der Spiegel* in a recent edition included an article titled 'The Gold of the new Pharaohs'. In that article the new developments are being described and it is noted – with alarm – that Dubai would like to become the most modern city in the world, a metropolis which will merge East with West – Dubai as a mixture of Singapore, Seattle, and Saint Tropez. (*Spiegel Special*, February, 2003, pp. 126–129).

REFERENCES

AlSayyad, N. (2001) (ed.) *Hybrid Urbanism: On the Identity Discourse and the Built Environment.* Westport, Connecticut: Praeger.

AlShafieei, S. (1997) The Spatial Implications of Urban Land Policies in Dubai City. Unpublished Report, Dubai Municipality.

Belbin, C. (2002) A personal piece of paradise. *Identity*, November/December, p. 30

Chew, K. and Schmitt, U. (2000). *1001 Arabian Nights at the Burj Al-Arab.* Cyprus: ABC Millennium.

Elsheshtawy, Y. and Abul Nagga, M. (2000). Sustainable urban development in an age of globalization: towards a co-existence model in Dubai, UAE. *Proceedings of IAPS: Cities and Sustainability.* Paris: International Association for People-Environment Studies.

Friedman, J. (1986) The world city hypotheses. *Development and Change*, 17(1), pp. 69–83.

Gabriel, E. (1987) *The Dubai Handbook.* Ahrensburg: Institute for Applied Economic Geography

Ghannam, F. (1997) Re-imagining the global: relocation and local identities in Cairo, in Oncu, A. and Weyland, P. (eds.), *Space, Culture and Power: New Identities in Globalizing Cities.* London: Zed Books, pp. 119–139.

Heards-Bey, F. (1982) *From Trucial States to United Arab Emirates.* Harlow: Longman.

Hirst, D. (2001) The emirs in the internet era: Dubai, a sheikhdom happy to embrace globalization. *Le Monde Diplomatique*, February (www.mondediplo.com).

Jacobs, J. (2001).*Edge of Empire: Postcolonialism and the City.* London: Routledge.

Kassab, S. (1997) On two conceptions of globalization: the debate around the reconstruction of Beirut, in Oncu, A. and Weyland, P. (eds.) *Space, Culture and Power: New Identities in Globalizing Cities.* London: Zed Books, pp. 42–55.

Kazim, A. (2000) *The United Arab Emirates A.D. 600 to*

the Present: A socio-discursive transformation in the Arabian Gulf. Dubai: Gulf Book Centre.

Khan, A. (2002) Dubai financial, transportation hub. *The Nation, Online Edition* (www.nation.com).

McBride, E. (2000) Burj Al-Arab. *Architecture*, **89**(8), 116–127.

Munif, A. (1989). *Cities of Salt*. New York: Vintage Books.

Oncu, A. and Weyland, P. (eds.) (1997) *Space, Culture and Power: New Identities in Globalizing Cities*. London: Zed Books.

Said, E. (1978) *Orientalism*. New York: Random House

Smalley, B. (2002) A tale of two cities: By 2005 Dubai will be able to boast another city within a city. *STARS Online*, (www.stars.com/art).

Smith, A. (1991). *Creating a World Economy: Merchant Capital, Colonialism, and World Trade, 1400–1852*. Boulder, CO: Westview Press.

ACKNOWLEDGEMENT

The author would like to acknowledge the partial support of the UAE University Research Council in carrying out this study (Research Grant #03-03-7-11/03).

Index

Note: Figures are indicated by *italic page numbers*, notes by suffix 'n[]'